The Command Decisions Series

Congested Airspace

The Command Decisions Series

Congested Airspace

A Pilot's Guide

Kevin Garrison

Belvoir Publications, Inc.
Greenwich, Connecticut

ISBN: 1-879620-13-8

Printed and bound in the United States of America by Arcata Graphics, Fairfield, Pennsylvania.

Contents

Introduction

Today's skies are crowded. No one will dispute that, particularly those who fly near the nation's major metropolitan areas. The congestion poses a problem for all pilots, and for many it's a nut so tough to crack that they simply stay away altogether.

For these pilots, the maze of airspace and procedures that comprise air traffic control can be confusing, frustrating and intimidating: something that's to be avoided unless absolutely necessary. Sooner or later, though, most pilots will need to venture into a busy airport or a major terminal area, and will be forced to work with ATC—and they may be ill-prepared to handle the experience. These pilots don't get as much out of flying as they could. Rather than using the system to their advantage, they avoid it.

But, it doesn't need to be so. Any pilot, provided he flies a suitably equipped airplane, can use ATC at any time to make his flying more efficient, expedient, safe and even enjoyable.

Even IFR pilots, much of whose training is in how to use the system, may not be fully aware of all of its features and how to take advantage of them. It's a matter of understanding how the system works from the inside out, and knowing how to approach it as a resource to be used to the pilot's benefit.

Small airplanes can be useful tools. They can get a pilot and his passengers to places that may be impractical or even impossible to reach via the airlines, at almost any time they choose. Using the tool properly requires more than just the ability to fly the airplane; it requires the pilot to be able to handle flying in a crowded terminal environment.

Entering the Hornet's Nest

Let's take a look at how an airplane can be used to great advantage. We'll take an imaginary flight with a low-time private pilot who goes out of his way to make the system work for him.

A 150-hour VFR pilot (let's call him "John Smith") had a problem not long ago, the kind that small airplanes are the ideal solution for. Smith, a businessman in Burlington, Vermont, had to get two clients together for a meeting in Glens Falls, New York, at the offices of one of them. The other client would be arriving in New York City from Atlanta the day before the meeting for other business. Afterwards, Smith and the traveling client would return to Burlington.

It was no problem for Smith to get to the meeting: Glens Falls is only 70 miles from Burlington. The trouble was getting his client from New York to Glens Falls. There was no scheduled airline service to the airport there, and the nearest alternate is Albany, 30 miles to the south. That would mean inconvenience for the client, who would have to get to the airport to catch a regional flight to Albany, then somehow get to Glens Falls—probably via a rental car which, of course, would then have to be returned. It was immediately obvious that there wouldn't be enough time to get everything done in one day.

Smith had a better idea. He'd fly to La Guardia himself, at a time convenient for the client, and take him straight to Glens Falls. Not only would that save time and hassle, the personal service would go a long way with the client.

Smith had never done anything like this before, but he was confident that he could handle the morass of traffic in the New York area. It would mean dealing with an intense, fast-paced ATC environment and a major airport at which he would definitely be a second-class citizen. But, Smith felt up to it. He was already fairly good on the radio: Burlington International is in the middle of an Airport Radar Service Area (ARSA), so he was comfortable with ATC chatter. He'd always made a habit of using all the publications available to him, so he knew where to go for information about the route.

He started planning three days in advance. He bought a fresh sectional chart, a New York TCA chart, Airport/Facility Directory, and just to keep the bases covered, an IFR enroute chart and book of approach plates. Not that he would (or could) go IFR, but he knew that charts for the IFR pilot have information on them that can be useful for VFR pilots.

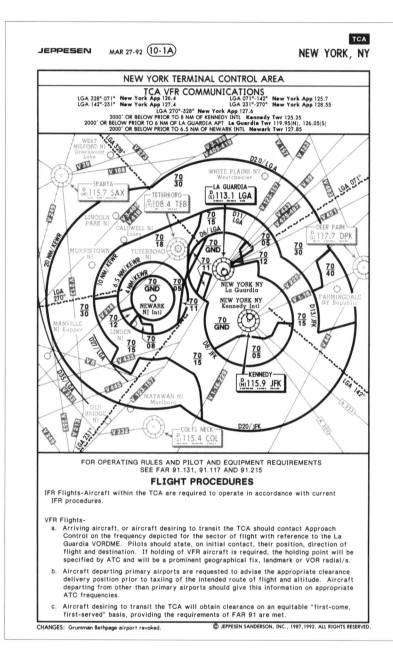

The New York TCA as depicted on Jeppesen's IFR area chart. Carmel VOR lies about 12 miles northeast of Westchester County Airport; La Guardia is roughly in the center of the TCA.

While at the FBO, he took a look at one of the commercial FBO guides, finding one at LGA where he could meet his client and get fuel. A call to them would give him the details on landing and parking fees.

Smith knew that the toughest part of the flight would be in and near the New York TCA, and he started by reviewing the charts so he would know what to expect when he got there.

The nearest VOR to the north is Carmel, so he would plan his route to pass over it, knowing that it would make a good reporting point for his initial call-up of New York Approach. By picking a VOR as a reporting point, he wouldn't have to worry about identifying landmarks in an unfamiliar area.

He got the approach control frequency from the TCA chart, along with several additional frequencies from the A/FD and the La Guardia approach plates. These would go on a list in the cockpit, along with all the frequencies he'd need en route. These he picked up from the IFR enroute chart.

The LGA approach plates also provided him with an excellent airport map with all taxiways labeled, along with the general aviation terminal.

Having collected all of the preliminary information about his destination, he proceeded to plan the route. He laid it out on both the sectional and IFR enroute charts, using a highlighter on the IFR chart to make it easier to follow. He elected to use airways for the trip and navigate primarily by VOR, keeping up with pilotage as well in case he got lost or if his radios failed.

His planned alternate, in case he was unable to get into the TCA for some reason, was White Plains (HPN), a reliever airport located under the northern edge of the TCA. There he could call his client and get fuel.

As a final preflight detail, he made reservations for a commuter flight to Albany in the client's name and set up a rental car in case the entire flight fell through.

Smith called his client and they arranged to meet at 9:30 a.m. at the FBO in the general aviation terminal at La Guardia.

The day before the flight, Smith started looking at the weather situation in detail, checking the television forecasts and using his computer to call DUAT. This gave him an overall picture, and let him do a preliminary check of notams. It looked like the weather would be okay, though possibly not: there was a cold front moving through the area, which was expected to have cleared out by the time of the flight. Behind the front were turbulence and high winds.

Before dawn on the day of the flight, Smith again logged on to DUAT for a full briefing, which he printed out and studied. The weather was to be as expected, which meant that he would have to fly at a lower altitude than planned to avoid unfavorable winds.

He figured out his time en route, checking to make sure there would be sufficient fuel reserves. He gathered his materials, including a filled-in flight plan form, and called Flight Service. He liked to get a "live" weather briefing as well as a computer printout, in case he missed anything. At the end of the briefing he filed his flight plan, and headed for the airport.

Before going out to his tiedown, he called the FBO at La Guardia with a message for the client giving his estimated time of arrival.

After preflight, Smith collected his materials: charts opened and folded so his route showed, approach plate book with LGA's plates marked with a rubber band, radio frequency list, flight log and blank paper for notes. He consciously set everything up so he wouldn't have to fumble for it: he was going to be busy enough looking for traffic and talking on the radio without extra hassles in the cockpit.

He picked up the ATIS, then called ground control for a departure clearance (Burlington has no clearance delivery frequency) and transponder code.

After takeoff, Smith contacted Burlington Approach as directed and asked for flight following. This early in the day, there wasn't much traffic to worry about, but it kept him hooked into the system and aware of what was going on in the area. After setting himself up for cruise flight, he took a moment to leave the frequency, contact Flight Service and activate his flight plan. He also picked up the latest weather at La Guardia. Noting that the winds were northerly, he decided it was likely that La Guardia would be using Runway 4. Another check of the TCA chart set him up mentally for the arrival.

As he left the area, the controller suggested another frequency to try for further advisories. Each controller along his route did this, so there was never any real need for him to refer to his list of frequencies.

Traffic began to pick up around Albany. He was not penetrating the Albany ARSA, but was talking to the approach controller nonetheless. Four commuter flights were called out for him: he saw none of them, but that wasn't unusual.

About 15 miles north of Carmel VOR, Smith began cleaning up and preparing for the entry into the New York TCA. He canceled flight following, then called Flight Service and canceled his flight plan. This eliminated one more thing to worry about; once he was in contact with New York Approach it would have no real purpose anyway. The sectional and IFR enroute charts were stowed. The TCA chart was put in a handy spot, as was the approach plate book, open to the expected approach (ILS Runway 4). This provided him with a handy reference for several navaid frequencies, including the locator-outer marker and La Guardia and Kennedy VORs.

He set his radios up. Number 1 nav was already on Carmel. Number 2 was on La Guardia. The ADF was tuned to the outer marker frequency. Number 1 comm was on the La Guardia ATIS frequency, and Number 2 comm was on the approach frequency printed on the TCA chart.

He picked up the current La Guardia ATIS—he'd been right about the approach—and noted a different frequency than the one printed on the chart was being used for approach. He tuned this in, and wrote it down for reference.

The approach frequency was busier than he expected. It seemed the controller was trying to talk as fast as he could, and the pilots were responding so fast he wondered if he would be able to get a word in edgewise. The controller was talking so fast it was sometimes hard to make out every word, but Smith used a trick his instructor taught him about radio communications: the controller will always say the same sorts of things in the same order, so in a sense you already know what he'll say before he says it—you know what to expect.

He listened for a minute, picking up the flow of communications. He made his call the moment the controller stopped talking, and used another trick—rather than go through the entire routine of who, where, and what you want, he compressed his transmission into a single phrase:

"La Guardia Approach, Cessna 1234 over Carmel with a request."

The controller gave instructions to three more airplanes, then said, *"Cessna 1234, stand by."*

Smith reduced power; he didn't want to penetrate the TCA while waiting for a clearance, and he was now only ten miles from its edge. The controller issued another high-speed stream of orders to four more airplanes, then said, *"Cessna 1234, go ahead with your request."*

"Approach, Cessna 1234 over Carmel with ATIS Yankee, landing La Guardia."

"Cessna 1234, squawk 5346 and ident, say type Cessna."

"5346 on the transponder, and we're a 182."

More orders to more airplanes, and the controller said, *"Cessna 1234, radar contact three south of the Carmel VOR, descend and maintain three thousand, fly heading 200 for vectors to the left downwind, Runway 4."*

Smith read back the instructions while scribbling them on his pad, then checked his TCA chart. It appeared the controller was going to take him around the Westchester ATA, and down the Hudson River, then cut him across the northern tip of Manhattan for the pattern entry.

The radio traffic never stopped. Three targets were called out for him. He switched on his landing light, the better to be seen. He saw several more airplanes, astonishingly close to him, but none were an actual threat. There was one frequency change on the way into the airport. That controller turned him towards the field and told him to report it in sight.

It took a moment in the urban sprawl, but he could clearly see the jetliners landing and departing and saw it in a moment, perched on the edge of the water, with runways literally built on piers.

He was kept up at 2,500 feet on downwind, and told repeatedly to keep his speed up. He was handed off to the tower controller, who put him between two DC-10s on final. He was admonished again to make his best speed, and cautioned about wake turbulence.

Smith was at cruise, and had been left far too high. He hadn't really expected this, and slipped the Cessna to get a high sink rate without building forward speed. Even so, he crossed the numbers at about 120, and wound up floating for what seemed like forever, but was actually about 2,000 feet. The controller told him to get off the runway immediately because of the DC-10 on short final.

As Smith turned off, he looked back in time to see the enormous jet about to touch down—too close for his comfort, but at least he was clear.

The ground controller was harried and abrupt, and Smith was glad he'd done his homework and didn't need taxi directions.

As Smith sat with his client while the airplane was fueled, he reflected on what he'd just experienced. It was really hard work trying to keep up, and he'd actually broken a sweat in the cockpit, feeling like a one-armed paper hanger. He realized that the only thing that had gotten him through it all successfully was that he had no illusions about it, and had taken every step to find out what to expect ahead of time. It had all paid off.

Our imaginary pilot did something that few VFR pilots would ever consider doing. He did it in the right way, too, through thorough preparation, ability to change plans to suit circumstances, and staying on top of the situation at all times.

What We're All About

Too often, we hear nervous VFR pilots trying to make their way through a TCA without really knowing what's going on. Almost invariably, the overworked controllers get very terse with them and, in effect, tell them to get the heck out of their hair. Frankly, one can't blame them. They have better things to do than try to nursemaid a pilot who doesn't really know how to conduct himself inside that crowded chunk of sky.

And that, in part, is what this book is all about: to give pilots an in-depth look at how to get into, through, and out of the most crowded airspace in the country.

We hope to pass on some of the finer points of dealing with the system for experienced pilots who want to make more effective use of it. In addition, we'll provide a solid basis for those VFR pilots who want to get their feet wet, possibly with the goal of obtaining an instrument rating in the future. Even if an IFR ticket isn't in your plans, the information found here can greatly enhance your flying from a safety, skill and enjoyment standpoint.

We've endeavored to approach the system with the goal of giving the pilot the tools necessary to make the best possible use of it, to get on top of it and make it work for you.

There's a lot here that can only be learned if you're on the inside, knowledge that may never be taught in a formal training program.

There is advice from those who have been there, pilots who routinely fly into the busiest airports in the country. We'll cover specific procedures, hints and tips for getting into and out of every major airport in the United States.

The new airspace layout to go into effect in the fall of 1993 will be covered in detail along with the existing airspace, to make the transition easier for pilots used to today's alphabet soup of ARSAs, TCAs, ATAs, and so forth.

We'll show where ATC came from, where it's going, and how it affects you, the pilot. We'll give you solid, sound advice on how to improve your communications and organization skills to make better use of ATC.

The entire process of a flight that makes maximum use of the system will be covered, from planning to arrival, for all sizes of airport, from uncontrolled strips to major metropolitan hubs.

Lastly, we'll provide extensive coverage of safety tips for pilots flying in congested airspace, and how to make use of ATC to help you out of an emergency.

Your Tax Dollars at Work

The system is one service of the Federal government that benefits all those who fly, whether they make direct use of it or not.

It's in the best interests of the pilot to seize the opportunity provided by this free service, for it has much to offer. It's our goal, with this volume of the *Command Decisions* series, to encourage that use, and to give you, the pilot, the knowledge needed to get the most out of the experience.

Not only will you find the undertaking rewarding, you may well find that it's enjoyable, too.

ir traffic control as we know it today can be traced directly back to a single event: the mid-air collision of two airliners over the Grand Canyon in 1956.

That's not to say that there was no means of controlling the flow of air traffic before then—many of the elements of ATC were already in place—but today's elaborate system of navigational aids and radar-based traffic control was developed as a response to the Grand Canyon disaster. That crash also resulted in the formation of the FAA (originally the Federal Aviation Agency, now the Federal Aviation Administration).

The overriding purpose of air traffic control is to keep what happened over the Grand Canyon from happening again: mid-air collisions. A secondary purpose—expediting traffic—is also part of its mission. All of the rules, controllers, and facilities exist essentially to keep airplanes from running into one another. Other parts of the system are devoted to other tasks. For example, Flight Service is there to provide weather information, and the network of navaids is there to get pilots where they're going. All of the instrument landing systems and published approaches which at first glance seem intimately tied to ATC are really just there to get airplanes to the airport without making an unplanned arrival at ground level—if there were no other airplanes, a pilot could shoot ILSes all day without ever talking to ATC.

Though there are three distinct functions performed by "the system" (collision avoidance, weather information, navigational guidance), many pilots tend to think of it all as being run by ATC (how often have we all heard pilots asking approach controllers about weather?). This is understandable, since air traffic controllers are the people pilots talk to most of the time. But, it wasn't always so. In years past, pilots were much more concerned with actually getting to where they were going than they were about dealing with controllers.

Structure of ATC

To help understand how we got where we are, it's useful to look in broad terms at how things work today.

ATC revolves around controllers, each of whom is responsible for the aircraft operating in his or her airspace. The controllers work at facilities that are dedicated to a particular kind of airspace. For example, tower controllers are in the tower cab and are responsible for controlling only the traffic in the immediate vicinity of the airport. Approach and departure controllers work at or near the airport at TRACON (Terminal RAdar CONtrol), and are responsible for aircraft approaching or leaving the area of the airport, and for low-altitude traffic passing through the area. Center controllers are responsible for enroute traffic flying between terminal areas, and for high-altitude traffic overflying them.

Each level of controlling facility "works" a larger volume of airspace, and is thus a bigger operation. Who a pilot talks to depends on which chunk of airspace he's in. For example, a VFR pilot seeking flight following is probably too low to be in Center airspace, so the best choice for a frequency to call on would be the approach control frequency for the nearest airport. If the pilot guesses wrong, the controller will steer him or her in the right direction.

On a given IFR flight, the tower controller will "hand off" the pilot to the departure controller, who will in turn pass control to a center controller, who will in turn hand the pilot off to the approach controller on the other end, who sends the pilot to the tower controller. There may be several controllers at each level as the pilot flies through different sectors of airspace.

Also, many GA pilots in congested areas never even get into Center airspace—they're simply passed from one approach control to another as they leapfrog across the landscape. Over most of the country, a pilot can be in radar and voice contact with an air traffic controller from the moment he takes off to the moment he lands hundreds of miles away.

As recently as the early and mid-60s, though, control was much more spotty. And in the earlier days, it was non-existent.

Early Days

It took several decades for any real form of ATC to evolve in large part because there was little need for it in the early years of aviation. Weather, navigation and mechanical failure were much larger sources of danger than traffic conflicts. For this reason the earliest parts of what could be called an ATC system were navigational aids; the more immediate problem was attacked first.

Nonetheless, there was official concern for traffic control (specifically, collision avoidance) remarkably early on. Just after World War I the International Committee on Air Navigation was formed. Although few people remember or have even heard of the existence of this committee, their aviation lives are affected every day by the actions the group took; things like which side of the airplane red and green navigation lights would be on and right-of-way rules that we still use today. In the years that followed many countries, including the United States, took up the new rules of airmanship.

At about the same time there was a short-lived boom in aviation in the United States due to the return from overseas of experienced and unemployed military airmen and the availability of cheap, inexpensive war surplus aircraft. While most of this boom was expressed in "stunt" flying and airshows of all types, including "flying circuses" some of the returning airmen were more interested in more practical pursuits: airmail and later, passenger service. For the most part, though, the flying game was perceived as one of freedom and high adventure for the participant.

This changed somewhat in the late 1920s. An important clue to the future of aviation worldwide was the holding of a meeting in Paris, France that was known as the Convention of Paris. This convention laid down rules for European air traffic. Rules that were very much like those in Europe were adopted by the United States in 1928 shortly after the Havana Air Convention.

While these rules were resented by many who were concerned with the "freedom of the skies," they did set a legal precedent that would be important later. The new rules were very much like the laws regulating other forms of transportation found in the countries and in adopting them the nations clearly claimed sovereignty over the airspace above their territory: "air rights" as it were.

Although the first actual scheduled airline flight took place in 1914 with the flight of a Benoist Type XIV flying boat across Tampa bay from St. Petersburg to Tampa, official government concern over the safety and separation of air traffic began in May 1918 with the institution of the first airmail route between Washington, Philadelphia and New York. Routes from New York to Cleveland and Chicago soon followed.

In 1921, the first coast-to-coast airmail service was begun and with it the airline industry we know today.

Early Navaids

A much larger problem for the early airliners than collision was navigation, especially at night and in poor weather conditions. Prior to the pioneering work that Jimmy Doolittle did in blind flying in the late 1920s, all flying, including airmail, was done under visual conditions.

One of the first attempts to help the airmail pilot navigate his route at night was in 1921 when Jack Knight became famous for flying through the dark night to Chicago guided by bonfires set by farmers along his route.

Rotating beacons were the next logical step for night-time air navigation. The seventy-two mile stretch from Columbus to Dayton, Ohio was equipped with a series of rotating beacons. Army pilots made over twenty-five successful runs using these primitive navaids. In addition to the beacons, field flood lights and flashing markers were set along the course. The aircraft, de Havilland DH-4Bs, were outfitted with landing lights and two parachute flares.

This system of flares, lights and beacons was adopted nationwide and was in place by late 1927. More than 2,000 miles of lighted airways existed in the United States by the end of that year and they stretched from coast-to-coast.

No light, beacon or flare in the world could solve the problem the early pilots ran into frequently when trying to keep a schedule...bad weather, then as now, was a very large problem.

With the use of the newly invented artificial horizon, the directional gyro and the improvements made to the sensitive altimeter Jimmy Doolittle changed the aviation world by conducting the first completely blind flight in September 1929.

These new advances in navigation combined with the improvements made in the design and reliability of the aircraft of the day gradually led to the problem of air traffic. With the establishment of airline passenger service in 1927, the advent of the first truly "modern" airliners; the Boeing 247 and the DC-2, a new way of travel and a new problem was born.

The nation's airways, of course, connected the large cities and in so doing concentrated all the air traffic over the areas with the highest populations. While no one wanted to think what would happen if two aircraft would collide and crash over an area like Manhattan, they had to admit that now the possibility did exist.

Early attempts at air traffic control began at the busier airports found at the large cities. The signals employed by the "ancient controllers" were all visual in nature and of course were useless in low visibility conditions and at night.

To control the traffic pattern a person placed himself in a prominent place on the field where the pilots had a good chance of seeing him and used two flags to signal approaching aircraft; a checkered or green flag to approve landing, a solid red one to hold or go-around. Other refinements that came along later were the light-gun signals (still in use today) and in the 1930s, radio. The first experimental radios were used in Cleveland in 1930. Others were installed at other locations afterward, using low powered transmitters. These methods were used mainly to control takeoffs and the pilots were under no real compunction to always follow the instructions.

Beginnings of Traffic Control

In 1935 the fear of collision became so compelling that American, United, TWA and Eastern Airlines set up a corporation and established an airway traffic control unit in the Newark airport terminal building. This unit was to control the flow of traffic through the very busy New York area. The unit was deemed a success and similar ones were set up in Cleveland and Chicago.

Although these facilities were successful, they only solved a tiny fraction of the air traffic problem. Private aircraft as well as military did not have to take part in the control program...there were no rules telling them to do so. In addition, the centers only controlled airliners in and out of the major terminals, had absolutely no direct communication with them and when push came to shove, had no real authority over them.

In 1936 the federal government took control of these airway traffic control centers and began their long journey into directing air traffic in the United States. Local airport towers still remained private.

The controllers at these early "centers" had to communicate with the pilots they controlled in a round-about fashion. When issuing clearances or instructions they first had to contact the aircraft's respective company. The airline dispatcher would then pass the information along to the pilot in the air either through direct radio contact or by relaying the message through Department of Commerce radio operators.

All separation of aircraft was accomplished by "dead reckoning."

Pilots would call in position reports at regular intervals over required locations, giving their estimate for the next fix much as IFR pilots do today in non-radar environments. The controller would then separate the aircraft by use of estimates and different altitudes. The controller's tools included a blackboard for flight progress, a map table and a teletype machine or telephone to relay clearances to the pilot.

It should be noted that control of traffic was only over established airways well into the late 1950s (area navigation, so common today in the form of RNAV, loran and GPS hadn't come along yet). If you were off the airways you were on your own and were completely out of "controlled" airspace. This not only meant that you could take any route you pleased but that the VFR separation and visibility minimums that we are so familiar with today did not exist. If you were able to fly on instruments you were perfectly legal to do so without a clearance of any kind. As a matter of fact, you can still do that today provided you can find what is left of uncontrolled airspace in this country (usually below either 1,200 or 700 feet AGL).

By the eve of the United States' participation in World War II there were fifteen full-time Air Route Traffic Control Centers (ARTCCs) that operated on a 24 hour basis. By the end of the war there were 19 centers (today there are 21) and most local control towers were operated by the Civil Aeronautics Administration. Things pretty much remained like that until the end of the second world war brought its tremendous changes in the system and the way people thought about air travel and airplanes.

After World War II

The end of the war brought new technology and ideas into the realm of air traffic control. The most obvious one, of course, was radar and the institution of direct radio communication between controllers and pilots. One of the underlying and less obvious ingredients of post-war controlling was the war-related experience brought home by Americans. For example, during the war American bomb raid planners had to accommodate flights of hundreds of aircraft at a single time. Another factor was that the pilots returning from the war to become corporate and airline pilots were accustomed to living with government control while they were in the air.

Even with the introduction of these new tools the system was slow to adopt them. It was late 1949 when the first two-way radio communications were instituted at Chicago center. It was a full six years before this capability was extended to the other centers (1955, the year before the Grand Canyon crash—it's obvious how relatively primitive the ATC system was at the time).

After the war the country was ill at ease with the idea of spending large sums of money for an upgraded air traffic control system. Although the Truman Administration had adopted a plan for modernization in 1949, the Korean War and a general public disinterest slowed funding and in 1955 the CAA's staffing level was the same it was in 1947 even though passenger traffic on the airlines had doubled since 1949.

A large part of the problem in post-war America was the need for long-range radars for the safe operation of air traffic. The Administration realized this need but was concerned with the cost. Unfortunately, it would take a disaster of epic proportions that could have been avoided by a modern ATC system to get the American public to demand safer skies.

In 1956 the midair collision of two airliners over the Grand Canyon alerted the American public to the shortfalls of the air traffic control system in the United States. As if rubbing salt into the wound, the accident didn't even happen in bad weather or in crowded skies. It happened in the clear above a cloud deck over a vast, uncrowded expanse of airspace.

Both airliners were totally legal, were following the regulations of the day and were operating in what they thought was a safe fashion.

The collision in question was between two four-engine transports that were operated by two major airlines with experienced pilots and good safety records. One was a Trans World Airlines Lockheed Super Constellation carrying sixty four passengers and a full crew of six and the other was a United Airlines DC-7 with fifty three passengers and a crew of five. Both were piloted by experienced senior captains and both were flying to the east coast from Los Angeles.

Remember that at the time airline communications were handled through middlemen. The TWA pilot was talking to his company dispatcher, and the United to his. At no time was the same controller talking to both flights. Also, at the time there was no radar coverage.

Initially, Los Angeles Center assigned the TWA Constellation to an altitude of 19,000 ft and the United DC-7 to 21,000. Because of some small build-ups in the Daggett California area, the TWA Captain requested a climb to 21,000 ft to avoid the clouds. This was denied because of the United flight at 21,000. Because positive control airspace didn't exist in 1956 the TWA crew was well within their rights to ask for VFR, 1,000 feet on top. The TWA crew requested this, got approval and climbed. As it happened, the tops were at 20,000 ft, putting the TWA flight at 21,000 ft. (in 1956 the "500-foot" VFR cruising altitude rule hadn't been written yet). They were not aware that there was a United flight at that altitude, nor was ATC required to tell them. The United crew was not told either. Remember, when VFR, traffic separation is your responsibility.

At 11:31 am, Los Angeles received the only verbal clue as to what was going on. It was a barely readable transmission from the United flight and only three words could be made out: "We are going..."

There were no survivors from the collision. For the first time, in horrifying detail the American public and the Congress realized just how primitive our ATC system was and how much improvement had to take place.

Because of the public outrage at the shape the nation's airspace was in money was, for once, not hard to come by for the improvement of air safety.

In 1958 Senator Mike Monroney of Oklahoma and thirty-three co-sponsors introduced a bill to create an independent Federal Aviation Agency "to provide for the safe and efficient use of the airspace by both civil and military operations and to provide for the regulation and promotion of civil aviation in such a manner as to best foster its development and safety." The bill was signed into law by President Eisenhower on August 23, 1958 and the FAA was born.

Two of the most important powers given to the fledgling FAA for the purposes of our discussion were:

• The control of the use of the country's navigable airspace and the regulation of both civil and military operations within that airspace in the interests of the safety and efficiency of both.

• The development and operation of a common system of air navigation and air traffic control for both civil and military aviation.

There was one huge surprise in the bill. In a country concerned with checks and balances and the monitoring of those in power, the bill provided that absolutely no one could overrule the FAA Administrator on decisions where safety was a factor. FAA, at first, even investigated its own accidents. The obvious conflict of interest led to the formation of the National Transportation Safety Board, an independent agency with no regulatory power of its own.

The '60s and '70s

It was fortunate that the FAA was in existence with enough power to regulate air traffic with the coming of the jet age. Shortly after the FAA's official life began the Boeing 707 took to the skies in scheduled service followed by many other jet airliners and later, corporate jet aircraft.

With the introduction of jets into the day-to-day life of the controller came new problems. No longer did the controller and the pilot have the luxury of a 400 knot closure speed with head-on traffic. Now, the closure speeds would be in excess of 1,100 knots.

In 1960 a huge building project was taken on by the FAA in order to upgrade and equip "an entirely new complement of air route control centers at thirty-two locations in the United States."

The system installed was a combination of computers and radar much like the semi-automatic ground environment (SAGE) system that the Air Force had been using at its Air Defense Command sites.

Also at this time more and more VORs were being put into service, which in went hand-in-hand with the advent of improved avionics. By the time the early '70s came along, avionics were beginning to get remarkably sophisticated.

Going into the 1970s, traffic congestion problems at the large terminals were addressed by the FAA's program of Terminal Control Areas. We'll discuss TCAs in depth in a later chapter.

According to the FAA the number of mid-air collisions and near misses dropped dramatically within months of the establishment of the TCA system.

The way ATC functioned matured during this time, and it continues to work in much the same way today, only with greater efficiency due to improved technology—better radars, computer systems and communications.

The combination of the computer and radar evolved throughout the 1960s and 1970s from the simple sets that could only detect primary targets and were cluttered with ground and weather returns to advanced presentations that included aircraft ID, speed and could predict traffic conflicts well in advance of them actually happening.

Continued advances brought digital radar, which improved detection of aircraft still further, albeit at the expense of weather depiction.

The '80s

From a personnel standpoint, the Air Traffic Control system's biggest crisis happened on August 3, 1981 when 11,438 controllers, represented by the PATCO (Professional Air Traffic Controller

Organization) went on an illegal strike and were subsequently fired by President Reagan.

The wisdom of the strike or even the mass firing of controllers is not pertinent here; an entire book could be written on that subject. It is important to note that for probably the first time the FAA had to face the fact that it had personnel problems that needed to be addressed. Some of these problems included controller "burn-out," heavy workloads with little rest, the lack of adequate communication with management and what was considered by controllers to be inadequate pay for the jobs they were doing. Many today would argue that the controller work force still faces many of these problems.

In the mid-'80s pilot groups became alarmed when FAA started increasing the number and size of TCAs and TRSAs, and developed the ARSA. It was viewed as an "airspace grab" and dire predictions were made about collisions occurring around the edges of TCAs as pilots tried to skirt them. In fact there were some accidents, including the tragic Cerritos, California mid-air, but by and large everything worked out fairly well.

The airspace continued to evolve, with major restructuring of the eastern U.S. (the Expanded East Coast Plan), but there haven't been any revolutionary developments since system as we know it was created.

Technology

The big news of the 1980s and 1990s has been improved technology. Some big ideas have had a lot of trouble, like the TCAS collision avoidance system. Originally, there were to have been three variants of it, and everybody was supposed to have one. The low-end version was dropped, and now it's only airliners that are equipped. The system is very late and not what it was supposed to be, but it still is a step forward.

Another boondoggle was the Mode S transponder, opposed by pilot organizations as a bit too "Big Brother" like. A Mode S transponder would send a unique code to ATC, so that they would know not only that there was an airplane there, but exactly who it was. The adoption of these units has been put on hold indefinitely.

Other high-technology gadgets have changed the way pilots fly, primarily in terms of navigation. The last five years has seen the advent first of inexpensive loran, then database lorans, and now global positioning system (GPS) receivers. These are now so inexpensive and so capable that virtually anyone who can afford an airplane can have one. There are even hand-held units complete with airport and navaid databases that renter pilots can carry with them to give previously unheard-of navigational capability in even the simplest airplanes.

Other avionics advances include thunderstorm detection equipment that costs a small fraction of what an airborne radar goes for, private collision-avoidance systems (not TCAS, but an alternative), moving map displays, full-bore electronic flight management systems for GA aircraft, even heads-up displays. It's now possible to equip a Skylane better than a front-line fighter of the 1960s.

FAA is scrambling to keep up with the onslaught. Recently the administration announced its commitment to develop GPS instrument approaches in 1993. New computer systems are being developed to replace the now-aging workstations at ARTCCs and TRACONs.

Ground-based technology has also advanced in the form of doppler weather radar, microburst warning systems, the newly developed ground control radar, and so forth—even prosaic systems like AWOS, which is nothing more than a robot weather station for airports without weather observers on duty. While outwardly the system may look much the same as it did 10 years ago, in reality it's far more capable and efficient than it ever has been.

The Future

Given the improvements in technology it's conceivable that there could be precision instrument approach procedures using GPS that have no ground-based equipment at all—no antennas or transmitters to maintain and calibrate. For that matter, it might even be possible to eliminate ground-based navaids altogether.

We're not in the prognostication business, but we will say that it's not likely. For the foreseeable future we'll still have air traffic control in much the same form as it appears today. There will be enhancements, for example the possibility of digital data transmission direct to an aircraft, but in the next century we'll still be talking to harried controllers and tuning in VORs.

Airspace

As this is being written, American pilots are still saddled with a surfeit of airspace types. Between TCAs, ARSAs, ATAs, control zones, transition areas, airways, jet routes, and the Continental Control Area and the Positive Control Area, things get pretty confusing.

Fortunately, the decision has been made to adopt the ICAOO (International Civil Aviation Organization, the same people who brought you the phonetic alphabet) international standards for airspace designation, which consist of a simple alphabetical series A through G. The closer to the beginning of the alphabet, the more restrictive the rules. Much simpler.

Even better, the new airspace corresponds to current airspace almost exactly, so the "geography" of our current airspace system remains largely unchanged. That's good news for pilots who are

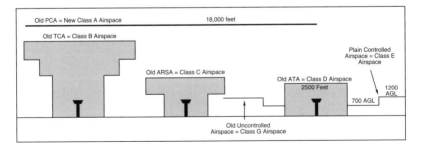

The new, far simpler ICAO airspace designation scheme.

used to the way things are now. Better still, some of the current airspace designations will become meaningless under the new system, so there will be less to remember.

Throughout our discussion of airspace and procedures, we'll endeavor to use both old and new designations to provide both a reference point and a guide for the way things will be starting in the autumn of 1993.

The Big Picture

We'll start by taking a very broad and simplifed look at airspace, then closely examine each type of airspace and what it means to the pilot. Most of the distinctions are really important only when operating VFR. When flying IFR, the pilot is hooked into the system and is under positive control for the vast majority of his time in the air, and therefore need not worry about restrictions or requirements: everything's already taken care of by his IFR clearance. (By the way, when the new airspace designations go into effect, the average IFR pilot probably won't notice anything different at all: there are no operating rule changes.)

Be aware that this first look is intentionally simplified. There are exceptions and oddities to much of what follows here which will be covered later.

The broadest category of airspace is that of *controlled* versus *uncontrolled* airspace. Every cubic foot of air over the U.S. is one or the other. Other designations may also apply: for example, the traffic pattern over a towered airport is not only controlled airspace, it's also inside an airport traffic area (Class D airspace)—but it's most useful to start with the idea of control and what it actually means.

In a sense, the very name is misleading. Flying in controlled airspace *does not* necessarily mean you're "controlled" by ATC. In fact, most of the time you're not. What flying in controlled airspace *does* mean is that the services of air traffic control are available to you, should you choose to use them.

In many of the more densely populated areas of the country, a pilot is almost always flying inside controlled airspace. In large areas of the west, by contrast, controlled airspace is relatively rare. In order to use ATC in those areas, it's necessary to stick to the established airways, or fly higher than most of us are equipped to do.

Uncontrolled airspace (Class G) is at the bottom of the ladder in terms of services available, rules imposed on the pilot, and standards required of him and his equipment. General controlled airspace (Class E—there is no U.S. equivalent for the ICAO Class F) is one notch above this. There are a few slightly more restrictive rules to obey and the pilot may use ATC if he so chooses.

The next step up is Class D airspace, which was known as the airport traffic area (ATA). This surrounds airports with operating control towers. Again, there are more rules to follow—contact with the control tower is required to enter, and pilots are required to follow the tower controller's instructions. There's an interesting twist here, in that the new Class D areas will actually be a hybrid of the existing ATA and control zone (more on this later on).

Class C airspace was called an ARSA. In keeping with the logical progression, there are more services available and more restrictions in force. Two-way radio contact is required before entering Class C/ARSA airspace, as is a transponder. VFR pilots can still fly pretty much where they like, but must tell the controller beforehand what they'll be doing and have to follow any specific instructions the controller issues. Traffic will be called out for VFR pilots, but separation is only provided between IFR aircraft and traffic that might conflict with them, not for VFR aircraft.

Class B airspace was known as a TCA. Again, there are more restrictions and requirements to be met. This is as intense and restrictive as it gets for most general aviation pilots. Pilots need permission to enter (this is, of course, automatic in the case of a pilot entering the TCA/Class B space on an IFR clearance), and, once inside, are under positive control of ATC at all times. If you don't do what the controller tells you to do, you're busted—and rest assured you *will* be getting specific instructions. What you get for submitting to ATC's control is positive separation from all other traffic, whether you're flying IFR or not. What you have to supply is two-way radio contact, a transponder with altitude reporting (Mode C), and a private pilot's license if you want to take off from or land at an airport inside the area. No student touch-and-go practice, please.

In effect, everybody flying inside a TCA/Class B area is operating just as if they were flying IFR. Of course, VFR pilots have to stay out of the clouds, and when push (the controller) comes to shove (the cloud), the pilot has to tell the controller he can't follow the instructions given him.

Class A airspace was the Positive Control Area, which concerns relatively few GA pilots. It extends from 18,000 feet all the way up to flight level 600 (60,000 feet), which is right up there in SR-71 territory. If you want to fly up there, you have to go IFR. Period.

As noted above, the majority of airspace in most of the country is Class E, plain-vanilla controlled airspace. It's like a blanket that overlies the Class G, uncontrolled airspace. This "blanket" is punctuated by thousands of Class D/ATAs, and well over a hundred Class C/ARSAs and Class B/TCAs. Transition areas and control zones, little pockets of Class E airspace that drop down into the Class G airspace below, exist around airports with instrument approaches. Above everything lies the Class A airspace.

Other Airspace Features

The above comprises the bare bones of the airspace structure. Other features, such as airways and jet routes, military operations areas and restricted areas, run through it as well, but each of these co-exists with one of the classes of airspace described above. For example, an airway is not a distinct kind of airspace; rather, it runs through the airspace known as Class E.

Note that Class G, uncontrolled airspace has no other features associated with it (there are a few exceptions); the space around an airway, for instance, or inside a Class B/TCA must by definition be controlled airspace. This makes sense: these features are all part of the function of air traffic control, which is not available in Class G airspace.

There are also control zones and transition areas, which are extensions of controlled airspace (Class E) that stick down below into the uncontrolled airspace below. The difference between them is how far down they go. The control zone as such will disappear when the ICAO system goes into effect, being combined with the ATA to make Class D airspace.

The final airspace feature is the Continental Control Area, which extends from 14,500 feet up to the floor of Class A/PCA airspace. It will vanish as a separate entity when the ICAO structure goes into effect. Basically, it exists to provide for controlled airspace features in those parts of the country that have none.

Class G/Uncontrolled—Class E/Controlled Airspace

Let's take a more detailed look at the various features of America's airspace, starting from the bottom and working up. Again, we'll start with controlled versus uncontrolled airspace, and look at the subject from a different angle.

The FAA's definition of controlled airspace seems to be a vague description of a complex subject. It really is vague because the subject is so complex. Let's look at the FAA's definition and then paraphrase it a bit to make it a little more understandable to the layman.

Class G, E Airspace Summary

Basic VFR Minimums		
Class G: Day, below 1,200 AGL: Clear of clouds, vis. 1 mi. Day, above 1,200 AGL: 500 below, 1,000 above, 2,000 horiz. from clouds, vis. 1 mi. Exceptions: FAR 91.155 (ops near airports - see VFR miminums chart)	Class E: 500 below, 1,000 above, 2,000 horiz. from cloud, vis. 3 mi.	Both Classes: Night: 500 below, 1,000 above, 2,000 horiz. from clouds, vis. 3 mi. Above 10,000 feet MSL: 1,000 below, 1,000 above, 1 mi. horiz. from cloud, visibility 1 mi. (Above 1,200 ft. AGL)
Was Called	**ATC Clearance Required**	**Equipment Required**
Class G: Uncontrolled Class E: Controlled	Class G: None Class E: For IFR	Class G: None Class E: Appropriate for IFR
Comm Required	**Separation**	**Services**
Class G: No Class E: For IFR only.	Class G: None Class E: For IFR and Special VFR only.	Traffic advisories and assistance on a workload permitting basis.

Controlled Airspace: Airspace designated as a continental control area, control area, control zone, terminal control area, transition area or positive control area, *within which some or all aircraft may be subject to air traffic control.*

Throughout the remainder of the definition the FAA then goes on to explain what they mean by a control zone, transition area and so on. The definition, although short and sweet, raises more questions than it answers. What do they mean by "all or some aircraft may be subject to air traffic control?"

The International Civil Aviation Organization's (ICAO) definition of controlled airspace makes a little more sense, is shorter and is more to the point:

(ICAO) Controlled Airspace: Airspace of defined dimensions within which air traffic control service is provided to controlled flights.

The general rule is that controlled airspace begins at 1,200 ft AGL. In a transition area, it goes down to 700 ft. AGL, and in an ATA/Class D area or control zone it goes down to the surface.

For our purposes, a combination and slight alteration of both definitions is a good idea. Why? Because controlled airspace means vastly different things to different operators, depending on what they intend to do in it. A control zone means something quite different to a 727 captain operating on an IFR clearance than it does to a 172 driver trying to maintain legal VFR and still operate in that airspace. Although they are both operating in the same airspace and trying to land at the same airport, different rules apply to these pilots and how they conduct their flights.

A basic misunderstanding among many pilots is that there are really many kinds of controlled airspace, each different. In a sense that's true, but there's a simpler way to look at it. True, the airspace inside a Class B/TCA is different from generic Class E/controlled airspace, but they're both controlled airspace. They share the basic feature of controlled airspace, which is the availability of ATC, and the basic requirement, which is to maintain VFR minimums. The real difference is that more restrictions apply inside the Class B/TCA.

In short, Class A, B, C and D airspace are all just like Class E/controlled airspace, only with more rules in effect.

What's the function of controlled airspace? Controlled airspace of all types, from Class B/TCAs to local Class D/ATAs exists for two simple reasons. They are, in order:

1. To exclude certain aircraft from that airspace for reasons of either poor weather or high traffic conditions for the purpose of providing the aircraft that do operate in that airspace separation and traffic avoidance.

This part of the definition would especially include such animals as the Class B/TCA and Class C/ARSA which will be covered in more depth later on. Even when the weather is CAVU, (ceiling absolute, visibility unlimited) many aircraft are excluded from this airspace because of equipment, pilot experience and licensing requirements. A non-radio Aeronca Champ flown by a student pilot would not be welcome in the traffic pattern at Atlanta's Hartsfield mixing it up with the L-1011s.

This function of controlled airspace has another important implication. In our simplified view of controlled airspace at the beginning of the chapter, we said that ATC was available in controlled airspace *should you choose to use it*. While a handy concept for the VFR pilot, it does *not* apply when the weather isn't good enough to fly VFR. When the conditions are too poor, VFR pilots are prohibited from flying in controlled airspace. This, in turn, implies that there are conditions when VFR flight in *un*controlled airspace is legal, but not in controlled airspace. And so it is—the minimums that apply are one of the major differences between controlled and uncontrolled airspace.

2. The second reason controlled airspace exists is to provide a framework in which other FARs can operate.

Many people don't realize that most of the VFR separation from clouds and visibility requirements are for *controlled airspace only*. Same thing applies for IFR pilots. Did you know that it is legal in this country to operate an aircraft in the clouds, IFR without a clearance in uncontrolled airspace? Yep, you can. All you need is an instrument rating and an aircraft capable of flying through the clouds. Remember, only in controlled airspace are the services of ATC available—and only in controlled airspace to the rules and restrictions that go along with ATC applicable.

The catch, of course, is that the uncontrolled airspace you'd be flying through would be so low and close to the ground, trees, buildings and power lines that under most circumstances you'd have to be crazy to try it, plus all the instrument approaches in the country are protected by controlled airspace and are legally unavailable in low weather unless you're under an IFR clearance—which you need to be in controlled airspace to have.

What Controlled Airspace Means to the VFR Pilot

To the VFR pilot flying in controlled airspace, the primary concern is maintaining basic VFR weather minimums. Much more restrictive visibility and cloud separation requirements exist when you're buzzing through controlled airspace than when you're operating outside the controlled airspace system. For example, in controlled airspace below 10,000 ft msl you must have at least three miles visibility. This way, on scuzzy, low visibility days when you want to fly out of a controlled airport you are excluded from the airspace. Why? To protect IFR flights operating in the area.

If you are at a non-towered airport that is outside other controlled airspace restrictions (transition areas, control zones etc.) you could operate the flight as long as you had one mile visibility and could stay clear of the clouds. The separation and visibility rules are familiar to every private pilot in the United States, or at least should be. The higher you fly the more separation and visibility you need in order to stay legal. The logic behind this is that the higher altitudes contain faster aircraft such as jets and your reaction time to traffic would be much shorter than at the lower altitudes. Below 10,000 ft msl, all traffic is limited to 250 kts IAS (indicated airspeed), above ten grand you can go as fast as you want and are able. In airliners like the DC-9 and 727 the pilots gently push over at ten thousand and accelerate to 300 knots or higher before continuing to climb. If you were VFR at 10,500 you would have a lot less time to see and avoid an airliner than you would at 2,000 where the jet would probably be slowed to 180 knots for approach with all its landing lights on.

The other important point about controlled airspace for the VFR pilot is the ability to call on ATC's services. A VFR pilot who wants to use flight following has to be in controlled airspace to do so.

What Controlled Airspace Means to the IFR Pilot

It's much simpler for the IFR pilot. Whenever you are operating under an ATC clearance you can pretty much assume that you are in controlled airspace and are afforded the protection of the VFR traffic being kept out of controlled airspace by the weather restrictions we mentioned earlier.

Controllers have absolutely no jurisdiction over uncontrolled airspace and can't issue you a clearance into it. They can cancel your IFR flight plan if you request them to but once you exit controlled airspace you are on your own. It should interest you to note that *there are no IFR restrictions in uncontrolled airspace.*

Can you fly through clouds in uncontrolled airspace without a clearance? Sure thing! Keep in mind though, how little uncontrolled airspace there is, how low to the ground it usually is and the fact that other idiots dumb enough to try this are probably around and you have absolutely no protection from them. Also, you cannot shoot an instrument approach out of this scenario unless you air-file because all instrument approaches are protected by controlled airspace.

Let's clarify this with a specific example. Imagine yourself on the ground at the Morehead, Rowan County airport just about to begin a flight.

Where does controlled airspace begin here? *Controlled airspace*

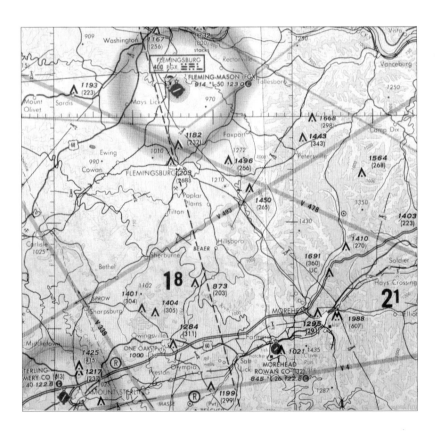

begins at exactly 1,200 ft above the ground. An important thing to note is that the floor of some controlled airspace is denoted in feet MSL (Mean Sea Level) while that of other controlled airspace is denoted in AGL (Above Ground Level). The floor of controlled airspace does begin at 1,200 ft AGL, but in high terrain, such as the Salt Lake City Utah area it is not unusual at all to be at an altitude of 5,000 ft MSL and still be in uncontrolled airspace.

Most student pilots would be able to tell you where controlled airspace begins above this airport. Now the important question: What does it mean? Can you take off VFR at will, or do you need to contact someone? It depends on only one thing: the weather. What about IFR? What happens from the time you take off until you reach the floor of controlled airspace?

What it means to a VFR pilot is that below 1,200 ft agl at this airport, he or she must remain clear of clouds with a mile visibility in order to be "legally" VFR. Above 1,200 ft the VFR pilot must have 500 below, 1,000 above, 2,000 to the side of any clouds and at least three miles visibility.

How do you tell if you're 2,000 ft to the *side* of a cloud? Well, it's really virtually impossible. There is no exact way to estimate your distance to the side of a cloud. It's also moot. Unless you have an FAA examiner on board with you, nobody's watching—and even if you did, he'd be guessing too. The visibility rule would have to be an estimate also unless you're operating near an airport that has a weather observer. What it really comes down to is give the clouds a wide berth. You never know when an airplane operating IFR might pop out of the side of that nice, puffy cumulus cloud, headed right at you. (By the way, that's why IFR flights are given whole altitudes, and VFR flights are supposed to stay at the 500-foot level between them.)

If you're operating an IFR flight out of Morehead, Rowan County you would not be afforded the protection of the ATC system even in bad weather until you were above 1,200 ft. If the weather were 500 overcast with one mile visibility there could still be aircraft flying in and out of this airport VFR legally and ATC has no control over them and probably couldn't even warn you about them. Since you're below the floor of controlled airspace, you are not protected by ATC.

Why is this? Why isn't there a transition area or CZ to protect you? Simply because Morehead has no instrument approach.

You can still make your IFR departure in dead rotten conditions if you like, and proceed on up to controlled airspace where you can latch onto the system. It's perfectly legal, though not prudent—after all, what happens if you're in the muck below controlled airspace, outside the system, and something goes wrong? The airport you just left has no approach, and you're essentially stuck.

Transition Areas

The first and simplest exception to this 1,200 ft AGL rule is shown to the north of Morehead, Rowan County Co. airport found on the Cincinnati sectional that we've been using for an example. It is Fleming-Mason airport.

There are three things you notice about this airport at first glance. First, there is a little parachute, denoting a jump zone...something important for you to keep in mind as you fly into the area. Second is the fact that there is a non-directional beacon on the field. This would tend to make you believe that there is probably some sort of NDB instrument approach to this airport. The third thing and the thing most important to our discussion is the fact that the airport is surrounded by a magenta-and-blue shaded circle. This is a transition area.

All that a transition area means, is that the floor of controlled airspace is lowered from 1,200 ft AGL to 700 ft AGL.

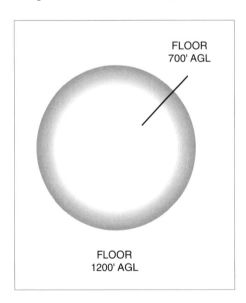

FLOOR
700' AGL

FLOOR
1200' AGL

An ICAO transition area: the circle is shaded magenta. On older charts there's a second blue-shaded circle outside the magenta one.

Usually a transition area is put over an airport or a group of airports that use a navaid such as a non-directional beacon or VOR to establish a non-precision approach. The main reason for the existence of transition areas is to exclude VFR traffic from the usual base of controlled airspace (1,200 ft AGL) down to 700 ft agl in order to protect that airspace for aircraft on an IFR clearance that are shooting the approach to that airport when the weather is bad. The minimum descent altitude (MDA) for the non-precision approach will be a bit higher than the floor of the transition area. IFR traffic is kept above the floor, and VFR traffic is kept below it—thus the approach is *protected* by being inside controlled airspace. Again, this *only* applies when the weather is below the controlled-airspace VFR minimums. On a clear day, the VFR pilot can ignore the distinction, and the only pilots shooting the approach are doing practice approaches anyway.

The concept of *protection* is an important one. While to a VFR pilot on a clear day controlled airspace means *availability* of ATC services, to an IFR pilot on a cruddy day controlled airspace is *protected* from intrusion by the erstwhile VFR pilot tempting fate.

If the transition area was not there it is possible that an IFR pilot on an approach could be outbound from the fix, in and out of the clouds with a mile visibility and be face to face with a VFR aircraft that was operating below the floor of controlled airspace. It is also possible using the same scenario, that a VFR pilot could run into an IFR aircraft flying through the clouds without a clearance below the floor.

If you are shooting an NDB approach into an airport like Fleming-Mason, you must keep in mind that the protection afforded by the ATC system ends when you descend below the floor of controlled airspace...in this case, 700 ft AGL. Once below 700 ft, traffic separation and avoidance is your problem.

If there are six aircraft in the VFR traffic pattern at Fleming-Mason when you break out it is your responsibility to avoid them and put yourself in the traffic pattern...they have a legal right to be there.

Transition areas will still be with us when the airspace designations change, but they'll be depicted differently on the sectional chart. Instead of a magenta and blue shading, there will be a single shaded magenta line where the floor of Class E/controlled airspace changes from 700 to 1,200 AGL—no more blue shading.

Minimums

A word on VFR minimums is in order, since they're the major difference between controlled and uncontrolled airspace from the point of view of the VFR pilot.

In Class G/uncontrolled airspace the only thing required to fly VFR during the day is a mile visibility and the ability to remain clear of clouds. These minimums are really awfully low for the average VFR pilot, and beg the question of whether you're flying on instruments or not. On a day with a mile visibility, you're unlikely to even see the edge of a cloud before you're in it, assuming the clouds even have well-defined edges in the first place. The closure rate for even the most modest of puddle-jumpers is over 200 knots, a distance

Basic VFR Weather Minimums*

Altitude	Flight Vis.	Dist. from clouds
1,200 AGL or less—		
Within controlled airspace	3 mi.	500 ft. below 1,000 ft. above 2,000 ft. horizontal
Outside controlled airspace		
Day (except as noted below)	1 mi.	Clear of clouds
Night (except as noted below)	3 mi.	500 ft. below 1,000 ft. above 2,000 ft. horizontal
>1,200 AGL, but <10,000 MSL		
Within controlled airspace	3 mi.	500 ft. below 1,000 ft. above 2,000 ft. horizontal
Outside controlled airspace		
Day	1 mi.	500 ft. below 1,000 ft. above 2,000 ft. horizontal
Night	3 mi.	500 ft. below 1,000 ft. above 2,000 ft. horizontal
> 1,200 AGL, and >10,000 MSL	5 mi.	1,000 ft. below 1,000 ft. above 2,000 ft. horizontal

The following operations may be conducted outside of controlled airspace below 1,200 AGL:
1. Helicopter. When visibility is less than 1 mi. during the day or 3 mi. at night, a helicopter may be operated clear of clouds if operated at a speed that allows the pilot adequate opportunity to see any air traffic or obstruction in time to avoid a collision.
2. Airplane. When the visibility is less than 3 mi. but not less than 1 mi. at night, an airplane may be operated clear of clouds if operated in an airport traffic pattern within 1/2 mile of the runway.

* An exception to these minimums exists inside Class B airspace, where the requirement is to remain clear of clouds with 3 miles visibility.

covered in 15 to 20 seconds. It's just plain unwise to fly in airspace that has the possibility of traffic under these conditions.

Of course, there are vast areas of the country where you can fly for hours at a couple of hundred feet AGL and never see a soul, and in that case flying close to VFR minimums in uncontrolled airspace is much safer. But in much of the country, flying when the weather is that poor is a bad idea because of potential traffic.

These low minimums used to apply at night as well, which was getting really close to the edge of safety. Just imagine trying to find an airport VFR at night when you can see only a mile. Recently FAA saw reason and increased the VFR night minimums to the same as for Class E/controlled airspace.

The minimums are also a bit higher above 1,200 AGL in uncontrolled airspace during the day. The visibility requirement is still only a mile, but Class E/controlled airspace cloud separation requirements apply.

In Class C/ARSA , D/ATA, and E/controlled airspace (and uncontrolled airspace at night) the minimums are 3 miles visibility, plus 500 feet below, 1,000 feet above, and 2,000 feet from the edge of clouds. As noted earlier, there are real problems in actually following this rule to the letter, but if you're close enough to the clouds in crowded airspace for the actual separation to be an issue, you're unsafe in any case.

These minimums apply below 10,000 ft MSL. Above 10,000, they go up to 5 miles visibility, and 1,000 below, 1,000 above, and a mile from the edge of clouds. The rationale here is that most traffic operating above 10,000 feet is traveling faster, and so more time is needed to see and avoid it. Even so, the minimums are still a bit close for real comfort, particularly in crowded skies.

The last twist to VFR minimums comes in Class B airspace. At present, basic VFR minimums apply inside a TCA. When TCAs turn into Class B airspace, the requirement will simply be to stay clear of clouds. This makes sense, since all aircraft are under ATC control inside a Class B area and are thus provided radar separation service.

Class D, Control Zones and Airport Traffic Areas

This is where the new airspace differs most from the old. First, let's describe the old system, then look at the new one to see how it's different.

Under the old rules, any airport with an operating control tower has an airport traffic area, or ATA. It extends out 5 *statute* miles from the airport in all directions, and goes up to 3,000 ft. AGL.

Class D Airspace Summary

Basic VFR Minimums		
500 below, 1,000 above, 2,000 horiz. from cloud, vis. 3 mi.		
Was Called	**ATC Clearance Required**	**Equipment Required**
Airport Traffic Control Area (ATA) Control Zone (CZ)	For IFR operations.	2-way radio. IFR - appropriate for the flight.
Comm Required	**Separation**	**Services**
Yes, with tower.	IFR, Special VFR, and runway operations.	Traffic advisories and assistance on a workload permitting basis. Control of traffic pattern.

Note that we said "operating" control tower. Many towers close at night, and when they do the ATA vanishes.

To get into an ATA, the pilot has to contact the tower. (This does not necessarily mean radio contact—if your radio is out, just make a phone call and arrange to use light gun signals.)

Inside the ATA, all traffic is under control of the tower. In effect, this means all aircraft in the traffic pattern—you'll rarely, if ever, hear a controller give specific directions to an airplane outside the traffic pattern other than to tell the pilot where to enter the pattern. When approaching the pattern from the edge of the ATA, a pilot operating VFR can do pretty much as he pleases: it's neither necessary nor appropriate to advise the tower of altitude or heading changes.

A control zone is different. It is often the same shape as the ATA (occasionally it has a "keyhole" extension), but has a different function entirely. All a control zone really means is that controlled airspace goes all the way to the ground in this area and all the way up to the continental control area.

A control zone, like an ATA, can be "switched off." Even though a CZ may be depicted, it may not be in effect. The usual requirement is that a qualified weather observer be on duty at the airport.

Whatever you do, don't confuse a control zone with an airport traffic area! If you're confused about the difference you're in good company. Although most control zones have airports in them that also have ATAs it is not a requirement. Also, control zones can have more than one airport in them. The point is that a CZ and an ATA *can* exist in the same place at the same time, but don't have to.

The most important difference between a control zone and an airport traffic control area is what is required of you, the pilot, in them.

In a control zone all that is required of you if you are a VFR pilot is that you maintain VFR weather and cloud separation minima that is required in controlled airspace. That's it—remember, a CZ is nothing more or less than controlled airspace. That does not obviate the need to listen to the tower controller if an ATA is in effect at the same time.

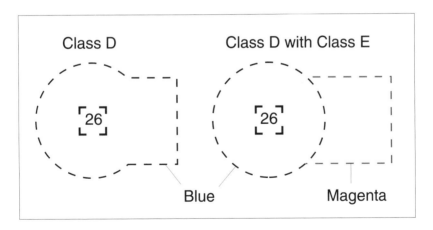

The biggest change on ICAO sectionals will be the Class D/ATA. Basically it will come in two varieties: the standard Class D area and the Class D area with an associated "keyhole" of Class E airspace that extends to the ground. The Class D area is outlined in blue, and the Class E in magenta. Note the area's "roof" altitude shown in hundreds of feet MSL.

In an ATA, you are operating on an actual clearance, either VFR or IFR issued from the tower and you are illegally in that airspace no matter what the weather without a clearance.

The airport traffic control area is never depicted on a chart. You know there is a tower there by the color of the airport on the VFR chart (blue). In the IFR world, you would know whether or not a tower was in operation at the airport by checking either the Airport/ Facility Directory during preflight, the enroute chart or the approach chart for that airport.

The control zone is depicted by a dashed blue line around the airport on the sectional chart. Some CZs are depicted by a line of blue Ts, which means that inside that CZ Special VFR flight is prohibited.

Under the ICAO system, the CZ and ATA will be combined to make Class D airspace. This not only simplifies things somewhat, it gives FAA considerable flexibility in designing airspace to suit the needs of a particular area. The old definition of an ATA dictated a fixed shape and size, which occasionally would mean that two airports close to one another would actually have overlapping ATAs, or that an uncontrolled field could be inside an the ATA of a nearby controlled airport.

With Class D airspace, the shape can be altered at will to allow for this kind of circumstance. A typical Class D area will have the keyhole-like extensions now associated with CZs to protect the space around instrument approaches, but overall may be smaller: the criteria for how far out to extend controlled airspace has changed along with the designation.

The other change is the ceiling: the old one of 3,000 feet is now down to 2,500 AGL. Also gone are all references to statute miles. Now, everything is measured in nautical miles.

Class D airspace will be depicted just as control zones are now, by a dashed blue line, often with a keyhole extension. It's possible to have a class D circle with a keyhole extension of Class E airspace that goes down to the ground (control zone). This will be shown as a dashed blue circle with a dashed magenta extension.

At airports with relatively small airspace extensions (no more than a couple of miles), the entire area will likely be Class D. At those with long extensions, the extension will probably be deemed Class E going down to the surface. This should keep pilots from "busting" Class D areas unwittingly—if a seven-mile extension were called Class D, a pilot could be far from the airport and breaking an airspace rule. On your sectional, just remember that you can fly through *magenta* dashed-line bordered areas without calling the tower, but be sure to contact them if you're about to penetrate a blue-bordered area.

One of the features of the charting scheme is that the upper limit of the Class D will now be printed on the chart in large, easy-to-read numerals. It's no longer necessary to squint at the tiny airport elevation note to figure out how high you must fly to stay clear.

One difference of note is the situation of a control zone around a non-towered airport. It really never was anything but controlled airspace that extended down to the surface anyway, and under ICAO that's what it will be called. This will simply be Class E airspace that extends to the ground. That portion of Class E airspace may well have effective hours much as CZs do now, dependent on a weather observer being on duty and communications being available.

It's worth noting here that although many of the old designations for various bits of controlled airspace are gone (e.g. Continental Control Area, transition area, control zone), the new system is burdened with a single designation and a series of caveats describing the same thing. For example, what was a CZ is now called Class E airspace that extends upward from the surface. What had been a transition area is not called Class E airspace that extends upward from 700 AGL. And what had been the CCA is now Class E airspace with a floor at 14,500 feet.

Of course, the various terms were really nothing more than labels for the different boundaries of Class E/controlled airspace, anyway, and were rarely referred to as distinct entities. On a given IFR flight, for example, you'd never even mention the CZ or transition area...and when was the last time you heard a VFR pilot talking about a control zone?

Class C/ARSA

The Class C/Airport Radar Service Area is the most likely environment in which you will find yourself trying to survive in the crowded skies of America.

Although the Class B/TCA gets the most attention in discussions about air traffic conditions, it is a fact of life that there are only 26 of these areas in the United States. There are presently 125 Class C/ARSAs active in the United States with more planned.

Basically, a Class C/ARSA is a formalized version of the way things worked near many airports anyway. Before the existence of the TRSA (Terminal Radar Service Area—a nearly-defunct voluntary version of the ARSA) and the ARSA it was common for VFR pilots to request radar service from Approach Controls and Centers.

Class C Airspace Summary

Basic VFR Minimums		
500 below, 1,000 above, 2,000 horiz. from cloud, vis. 3 mi.		
Was Called	**ATC Clearance Required**	**Equipment Required**
Airport Radar Service Area (ARSA)	For IFR operations.	2-way radio. Transponder. IFR - appropriate
Comm Required	**Separation**	**Services**
Yes, with ATC	IFR, Special VFR, and runway operations.	Traffic advisories and assistance. Control of traffic pattern.

The controllers would then provide traffic advisories and other services such as navigational help to the pilots who requested it on a workload-permitting basis.

The trouble with this approach was twofold. First, there was no standard way for the pilot and controller to do business: it varied from location to location and a pilot was never quite sure how to go about asking for help. Second, there was no formal requirement for the controller to help: the service was based solely on whether or not the controller's workload permitted him or her the time and the inclination. (Note that this is exactly the way radar advisories or "flight following" still works in Class E controlled airspace—on a workload-permitting basis.)

Another problem was that in the areas of high traffic volume where Class C/ARSAs now exist pilots could traverse the busy airspace without any contact at all with Approach Control or any controlling facility unless they entered an Airport Traffic Control

Area (ATA). Imagine flying along at 3,500 AGL right over the middle of the Orlando, Florida airport at the height of tourist season and you can see the problem; it was perfectly legal to do so without talking to either approach control or the tower, but it would put you very close to a large number of airliners and cause some mayhem at TRACON.

The intent and design of the Class C/ARSA solved these problems, recognizing the fact that it is important to provide radar to VFR as well as IFR pilots in congested areas. The Class C/ARSA also serves to standardize the operations of pilots in and out of these many areas and mandates that the controllers must have time to work with all traffic...not just on a workload permitting basis as in the past.

Dimensions Of Class C/ARSAs

There is no mystery as to the location and dimensions of Airport Radar Service Areas, they are spelled out very neatly in the FARs.

There is a basic, standard design that is changed only by specific, on-site variations. It's similar to the "upside-down wedding cake" of the Class B/TCA, only smaller.

The airspace of a Class C/ARSA is based on two circles, both centered on the ARSA's primary airport. The inner circle of the Class C/ARSA has a radius five nautical miles and has an upper limit of 4,000 ft AGL.

The outer circle has a radius of ten nautical miles. Its vertical dimensions begin at 1,200 AGL and top out even with the top of the inner circle at four thousand feet.

In addition to the Class C/ARSA there is an outer area, beginning at the ten nautical mile limit of the ARSA and extending twenty nautical miles. The outer area's bottom limit is the lower limit of radar/radio coverage for that area and extends upward to the upper limit of the respective Approach control's area. The outer area only has significance in that the same services given inside the Class C/ARSA are provided when radio and radar contact are established— pilots are not required to contact ATC, but can depend on service if they do.

ARSAs are shown on VFR charts such as sectionals by a segmented magenta line. They are also shown on some Terminal Control Area Charts.

When they become Class C airspace, the depiction will change to solid magenta circles, with the upper and lower altitude limits noted in both rings.

Just as with Class D airspace if there is a segment of Class E controlled airspace that goes down to the surface under the outer ring (what had been a CZ), there will be a dashed magenta line showing it.

Where are they on IFR enroute charts? Nowhere. Since you are operating with two-way radio communications with ATC anyway and are operating on an ATC clearance there really is no need for you as an IFR pilot to know exactly when and where you enter a Class C/ARSA's airspace.

As an IFR pilot, it would be a good idea to know when you are operating in a Class C/ARSA for the simple reason that you would know that you are getting just a little more traffic protection from ATC.

Primary/Secondary Airports

We should go over just what is meant by the term "primary airport" because it is used in both discussions of ARSAs and later in talking about TCAs. According to the FAA, the primary airport: "is the airport designated in Part 71, Subpart L, for which the airport radar service area is designated.

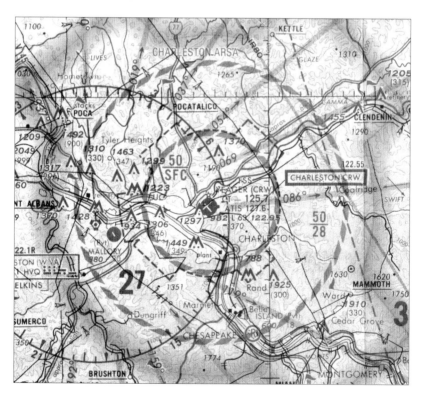

A satellite airport is any other airport within the airport radar service area." Huh?

What that really means is that when you look at any ARSA or ARSA chart the primary airport is usually the one the ARSA is named for. The Tallahassee ARSA's primary airport is Tallahassee Municipal Airport, for example.

What's the difference between a primary and a satellite airport? There are quite a few, beginning with the structure of the ARSA itself. Remember, the ARSA is designed and built around the primary airport. Even though a satellite airport may lie well within the ARSA, there is only one primary airfield per ARSA.

When departing from a Class C/ARSA's primary airport the rule is that no one may operate without maintaining two-way communication at all times with ATC.

At satellite airports, the rule according to the FAA is: "Aircraft departing satellite airports/heliports within the ARSA surface area shall establish two-way communication with ATC as soon as possible. Pilots must comply with approved FAA traffic patterns when departing these airports.

Obviously if the controlling facility (approach control) is at another airport miles away from the satellite airport it is usually physically impossible to establish two-way communications with ATC due to line of sight restrictions. Also, it wouldn't be very safe. It is far wiser when operating out of a small airport to be monitoring unicom than to be talking with a radar controller ten miles away that can't see you due to ground clutter anyway.

When you are arriving in a secondary airport that has its own control tower ATC will hand you off to that tower in plenty of time to establish communications with them. When the controller hands you off to the secondary tower's frequency its part of ARSA service is over.

Who Can Use a Class C/ARSA

There is no specific certification required for you to operate an aircraft in a Class C/Airport Radar Service Area. Pilots from the student pilot level up through the Airline Transport Pilot are perfectly welcome in a Class C/ARSA.

Remember, there is no requirement for an aircraft to be piloted by an instrument rated person due to the fact that a Class C/ARSA serves both IFR and VFR traffic. Especially VFR traffic because even without a Class C/ARSA, controllers are required to service IFR traffic.

Can a student pilot on a solo training flight operate in a Class C/ARSA? You bet. As long as the rules are followed that we will discuss a little later in the chapter, a student will get along fine and will have the added advantage of having "big brother" looking over their shoulder with traffic advisories and ready help such as vectors back to the airport if they get a little disoriented.

The professional pilot and the pilot operating on an IFR flight plan get the advantage of help in avoiding VFR traffic. This is especially valuable in marginal or scuzzy VFR conditions. Outside a Class C/ARSA, ATC does not have to provide this separation.

Arrivals And Overflights in Class C/ARSAs

Let's take a quick minute to see just what the FAA says about flying into or through a Class C/ARSA:

FAR Part 91.130: No person may operate an aircraft in an airport radar service area unless two-way radio communication is established with ATC prior to entering that area and is thereafter maintained with ATC while within that area.

You may have already noticed that nowhere in the FARs, AIM or even this book has anybody mentioned anything about a "clearance." All the rules say about flying into or through a Class C/ARSA is "two-way communications must be maintained with ATC."

Although ATC must be communicated with and they control your sequencing in and out of the ARSA's primary airport, they have no power to clear you to do anything that has to do with the ARSA. In other words, if you call them prior to entering the ARSA they cannot turn you back or deny you clearance through the ARSA...they can only deny you sequencing into the traffic pattern at the primary airport and issue you traffic warnings if you are VFR.

But, it's not quite that simple. There is a kind of "Catch-22" involved here. The FAR only says that you have to maintain two-way communication while in a Class C/ARSA and really doesn't say you must get a clearance to operate in one.

What happens, though when you call ATC up and they say: "Roger 123 yankee, maintain VFR at 3,500 feet and turn left to heading one three zero?"

In effect, even though you are VFR the controller has given you a clearance. Do they have the right to do that?

It is a kind of gray area in the regulations but it is covered by FAR part 91.123:

91.123 (a): When an ATC clearance has been obtained, no pilot in command may deviate from that clearance, except in an emergency, unless he obtains an amended clearance. However, except in positive controlled airspace, this paragraph does not prohibit him from canceling an IFR flight plan if he is operating in VFR weather conditions. If a Pilot is uncertain of the meaning of an ATC clearance, he shall immediately request clarification from ATC.

That paragraph makes it look like you only get a clearance from ATC when you are operating on an IFR flight plan and are in IFR weather doesn't it? The next paragraph is the kicker and is the gray area referred to earlier:

91.123 (b): Except in an emergency, no person may, in an area in which air traffic control is exercised, operate an aircraft contrary to an ATC instruction.

Well, a Class C/ARSA is certainly an area in which air traffic control is exercised, isn't it? So a basic clearance like an altitude and heading given to you when you are operating VFR in VFR weather is one you have to follow or be in violation of FARs.

Air traffic control certainly can not clear you to do something that is either against FARs or would cause you to have an emergency.

Let's say that that clearance ATC gave you when you initially called up would put you in the clouds. As a VFR pilot this is an emergency, right? Even if you are IFR rated and flying an aircraft capable of flying through those clouds, it would not be proper or legal to accept that clearance, because you're operating VFR and so must obey the VFR minimums for controlled airspace. Either it would be an emergency or a violation of the regs depending on your ratings and experience. Either way, you are not required to follow that clearance.

You *are* required to tell the controller that you are unable to follow his or her clearance and tell them why. Probably, they will then come up with another idea that will work.

If you follow a controller's clearance into a problem it is *your fault*. Actually, most of the regulations are set up to insure that whatever happens, it is your fault.

Air Traffic Control Services Provided

Air Traffic Control has four major responsibilities within the ARSA. The first and probably most important is the sequencing of all arriving and departing aircraft to and from the primary airport.

Since all ARSA primary airports are controlled by a control tower the requirement of sequencing arriving VFR and IFR aircraft is no biggie...the tower would have handled this anyway. It is to ATC's advantage, though, that they can pick-up and communicate with VFR aircraft a little further out (usually 20nm) and get them in line and advise them of traffic earlier.

The second requirement, that of maintaining IFR separation of IFR aircraft is something ATC does everywhere anyway. All the usual requirements of keeping IFR aircraft a certain distance and altitude away from each other applies in a Class C/ARSA just as it does in all other controlled airspace.

The third thing required of ATC in a Class C/ARSA is probably the most important to pilots, both IFR and VFR. Controllers are charged with the responsibility of separating both IFR and VFR aircraft within a Class C/ARSA. They will give traffic advisories and vectors to the conflicting aircraft so that their radar targets don't touch and at least 500 feet vertical separation is maintained. This gives you, the pilot just a little more protection than you would get in a non-ARSA environment. If you are IFR and are just about to break out of the clouds at three thousand feet you can rest just a little easier in a Class C/ARSA that you won't be surprised by a VFR aircraft just below the cloud bases.

Keep in mind, though, that 500 feet separation is not very much separation at all. Also, two radar targets "not touching" could mean a really close call.

The last requirement for ATC within the ARSA is providing traffic advisories and safety alerts to VFR aircraft. Notice that they don't have a requirement to maintain any kind of separation between VFR aircraft...only to advise you of the conflict.

Usually the controller will help you out if you are VFR and are converging on another VFR aircraft but is not charged with maintaining your distance from that traffic. If you both are VFR that is your job, and his.

The same services are provided by the controller in the outer area around a Class C/ARSA as long as two-way radio communication is established with the aircraft involved. Remember there is no requirement in the outer area to contact anybody and for that reason there may be a lot of traffic out there that isn't talking with ATC. The controller will try to point the traffic out to you anyway but won't have any information on its type or altitude.

If you are departing a Class C/ARSA and flying into the outer area ATC will assume that you want continued service into this area unless you tell them otherwise. Once outside the inner circle you can turn down or discontinue the service any time you want.

Also controller help in the outer circle may be shut off *by the controller* if their workload is too great for them to provide service to both IFR and VFR traffic. If that sort of workload arises the controller is obligated to serve the IFR traffic and you're on your own.

What Constitutes Two-Way Radio Communication?

It sounds like a trick question, doesn't it? Let's say that you are just about to enter a Class C/ARSA and try to contact the proper controller on the proper frequency. You call up giving your position, altitude and heading and the controller responds by saying, "Aircraft calling, please stand by."

Was that radio contact enough to enter the ARSA? According to the FAA two-way radio communication never took place because the controller never responded to your identification sign.

If the controller said, "Roger 123 yankee, please stand by", that *does* constitute communication and you can press on into the ARSA.

Is it really such a big deal, this worrying about poking your nose into a Class C/ARSA uninvited? Afraid so. Now days the controller is required to file a report when the violation is noted. Whether they actually do this or not depends usually on whether there was a traffic conflict and other factors such as whether or not a supervisor was standing behind the controller when the violation occurs.

When you are flying out of the primary airport of a Class C/ARSA making contact with the controlling facility is as easy as listening to the ATIS, calling clearance delivery or ground control for a transponder code and getting underway.

The tower controller will hand you off to the departure controller and you are more or less automatically "in the system."

If you happen to be flying out of a secondary airport in a Class C/ARSA getting in touch with ATC is a little harder but isn't too difficult. There are a few ways to get the proper frequency to get in touch with the controllers after you are airborne from the secondary airport. The first would be your good old sectional chart. For ARSAs there is a frequency box just outside the airspace depicted on the chart directing you to contact ATC on a discrete frequency.

Class B/Terminal Control Area

The last surface "animal" of controlled airspace before we begin our climb to higher altitudes is the Class B/Terminal Control Area.

Class B/TCAs are the "upside-down wedding cakes" that place

Class B Airspace Summary

Basic VFR Minimums		
Clear of cloud, vis. 3 mi.		
Was Called	**ATC Clearance Required**	**Equipment Required**
Terminal Control Area (TCA)	For all aircraft.	2-way radio. Mode-C Transponder (also under 30-mile "Mode C veil") IFR - appropriate
Comm Required	**Separation**	**Services**
Yes, with ATC	All	Traffic advisories and assistance. Positive ontrol of all airspace.

severe restrictions on aircraft operating in and out of their airspace. They surround all the large, high volume airports in this country and have been either lauded as the savior of Air Traffic Control or as yet another step in the government's curtailment of our flying freedom.

Used to be, back during the "good old days" of aviation you could fly your airplane just about anywhere in the country. Even if you had a sixty-five horse engine, no radio and rag wings you had every legal right and actually could land at places like Chicago's O'Hare or New York's Kennedy airports just about any time of the day you wanted.

Two new developments came up during the 1960s that would lead to changes in this policy. One was the increased traffic load at these places and the second was the advent of the jet airliner.

Both of these new twists in aviation history presented a problem that the flying public (and by that we mean non-pilot voters) demanded safer skies, especially around major terminal areas.

This public demand for more control combined with the new technology radars then available combined to begin the age of the Terminal Control Area.

After June 25, 1970 you could no longer fly your radio-free, transponderless airplane anywhere you wanted...the TCA became a fact of life at twenty-one of the airports in the United States.

Onerous over-regulation of our nation's skies? Maybe, but having TCAs around nowadays is simply something we Americans live with, like the IRS or all the taxes on fuel. As a matter of fact, after all these years since their inception, there are only a couple more TCAs today then there were in 1970. Before you get all excited, keep in mind that the airspace that was considered to be a "TCA Group III" is now designated as a Class C/ARSA and there are more than 120 of those around with more on the way.

Strangely enough, even though ARSAs outnumber TCAs by a factor of at least five, most pilots are really upset about TCAs and hardly ever mention ARSAs.

If you ask most pilots about flying into the Class C/ARSA in Des Moines and they will answer "no problem!" with a sly grin on their weathered faces. Mention flying into the Class B/TCA in Miami and they will dive under the table in fear.

The only real difference between the two kinds of airspace is that most of us have a lot of experience flying in and out of Class C/ARSAs and very little operating in and out of that "Debbil" Class B/TCA.

When the program first appeared there were three varieties of TCA (Group I, II and III), and TRSAs.

This has been simplified to TCAs and ARSAs, which will become Class B and Class C airspace. Not a bad deal, overall—two kinds of airspace instead of four.

According to the FAA: "A Terminal Control Area (TCA) consists of controlled airspace extending upward from the surface or higher to specified altitudes, within which **all aircraft** are subject to the operating rules and pilot and equipment requirements specified in FAR 91. Each TCA location includes at least one primary airport aaround which the TCA is located."

As usual, the FAA crammed ten minutes of information into a five-day seminar. What did all that in their definition of a Class B/TCA mean? Just this: If you are in a Class B/TCA no matter what you are flying from a lawn chair attached to weather balloons to a Boeing 747 SP you are required to follow certain rules.

That is easy because there are only 29 of them currently around.

Atlanta
Baltimore/Washington, D.C.
Boston
Charlotte
Chicago
Cleveland
Dallas
Denver
Detroit
Honolulu
Houston
Kansas City
Las Vegas
Los Angeles
Memphis
Miami
Minneapolis
New Orleans
New York
Philadelphia
Phoenix
Pittsburgh
Orlando
St. Louis
Salt Lake City
San Diego
San Francisco
Seattle
Tampa

Unlike Class C/ARSAs which are only depicted on VFR charts, Class B/TCAs are shown on Sectional, World Aeronautical, Enroute Low Altitude (an IFR chart), DOD FLIP (approach plates) and TCA charts.

A Class B/TCA is shown on the VFR charts as a series of solid blue lines formed in the shape of the airspace. This depiction will remain the same after the switch to ICAO airspace, except that the control zone depiction will disappear.

A look at the Atlanta TCA on the next page will show you two important differences between Class B/TCAs and Class C/ARSAs.

The first is that a Class B/TCA is much larger that a Class C/ARSA. While ARSAs can be of different sizes, they are rarely larger than twenty miles in diameter. Look at the Atlanta Class B/TCA; it is roughly seventy one nautical miles in diameter!

The second thing you will notice is that there are quite a few rings in this depiction of the TCA. They show not only the dimensions of that part of the TCA but give you the altitudes the TCA begins and ends. On the outer rings you will see:

<u>**125**</u>
100

This simply means that the Atlanta Class B/TCA's floor at this point is ten thousand feet MSL and its top is twelve thousand five hundred feet MSL. Don't make the mistake of using the figure you

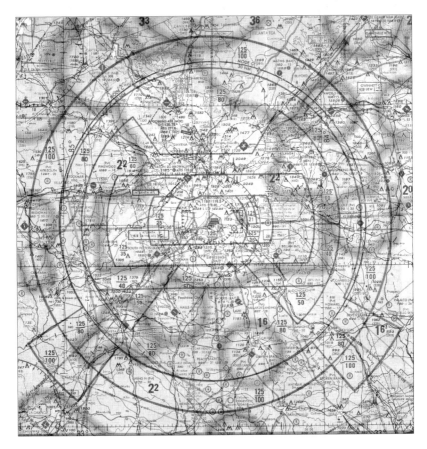

see on the south side of the TCA near the range marks. That large 1 with the smaller 6 near it, written in blue can be confusing. That figure has nothing to do with the TCA, it is a maximum terrain elevation figure. This appears in every lat/lon grid square on every sectional chart. NOAA takes the highest terrain feature in the area of the chart bounded by ticked lines of latitude and longitude, add a safety factor of either one hundred feet or one half the contour interval whichever is higher and express the figure as a two digit number, rounding it to the next highest hundred. In the case of the 16 shown on the Atlanta Sectional, the maximum terrain feature in that section is lower than one thousand six hundred feet MSL. Keep in mind that this figure does not take man-made obstructions into account.

On IFR enroute charts, TCAs are depicted as surrounded by a light blue border that is interspersed with white dots.

On the low altitude en route chart, just the outside edge of the TCA is shown. None of the altitudes are depicted. On the High Altitude enroute charts or "jet route" charts TCAs are not depicted at all. This is because while flying using a High Altitude chart you are operating above flight level 180 and are above any TCA airspace.

On Department of Defense FLIP charts and Jeppesen Approach Plates the TCA is shown on a separate TCA chart. The fact that you are operating in a Class B/TCA does not appear on the approach plate you are using for a specific approach, because it doesn't matter.

Look over the Boston Terminal Control Area plate 10-1A from Jeppesen on the next page. This is really a simpler presentation of what you would find on a VFR chart. The range circles are there as in the VFR version, the major airways and navigational fixes are shown and the floor and ceiling figures are all shown on the left hand side of the chart near the circle borders.

Notice also on the Boston chart that the dimensions as well as the floor and ceiling figures are quite different than those for Atlanta. In Atlanta the floor in the outside circle of the TCA is ten thousand feet. In Boston it is four thousand. The ceilings are different also. Boston's is seven thousand, Atlanta's is twelve thousand five hundred. This brings up a very important point about TCAs that is different from ARSAs: **Although all TCAs are generally similar, they are specifically distinct, or in other words; when operating in or around a Class B/TCA be very careful to read the small print.**

The last feature to look for regarding a TCA is the Mode C "veil," depicted as a similar but thinner blue line beyond the edge of the TCA. Within this area all aircraft must have an operating Mode C

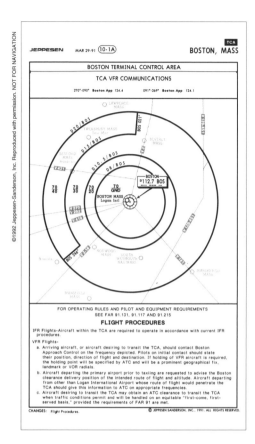

JEPPESEN   MAR 29-91 (10-1A)   TCA   BOSTON, MASS

BOSTON TERMINAL CONTROL AREA

TCA VFR COMMUNICATIONS

270°-090° Boston App 124.4   091°-269° Boston App 124.1

FOR OPERATING RULES AND PILOT AND EQUIPMENT REQUIREMENTS
SEE FAR 91.131, 91.117 AND 91.215

FLIGHT PROCEDURES

IFR Flights-Aircraft within the TCA are required to operate in accordance with current IFR procedures.

VFR Flights-

a. Arriving aircraft, or aircraft desiring to transit the TCA, should contact Boston Approach Control on the frequency depicted. Pilots on initial contact should state their position, direction of flight and destination. If holding of VFR aircraft is required, the holding point will be specified by ATC and will be a prominent geographical fix, landmark or VOR radials.

b. Aircraft departing the primary airport prior to taxiing are requested to advise the Boston clearance delivery position of the intended route of flight and altitude. Aircraft departing from other than Logan International Airport whose route of flight would penetrate the TCA should give this information to ATC on appropriate frequencies.

c. Aircraft desiring to transit the TCA may obtain an ATC clearance to transit the TCA when traffic conditions permit and will be handled on an equitable "first-come, first-served basis," provided the requirements of FAR 91 are met.

CHANGES: Flight Procedures.   © JEPPESEN SANDERSON, INC., 1991. ALL RIGHTS RESERVED.

(altitude encoding) transponder on board, though there is no requirement to contact ATC unless you want to enter the TCA.

Equipment Requirements And Operating Rules

The first and biggest rule for operating in any kind of TCA is this: **Regardless of weather conditions, ATC authorization is required prior to operating within a Class B/TCA.**

Unlike other airspace where your decisions about obtaining clearances and instructions from Air traffic Control were predicated on whether the weather conditions were VFR or IFR you are *always* operating on an ATC clearance when you are flying in a Class B/TCA.

No matter what you are flying, you will need the following to operate in a Class B/TCA:

• A two-way radio capable of communicating with ATC on appropriate frequencies. This for civilian pilots would usually mean some sort of VHF transceiver of at least 360 channels.

• A VOR or TACAN receiver, except for helicopters. No other navigational equipment is required. Any clearances you get from ATC while operating in a Class B/TCA must use VORs unless they know from your flight plan strip that you have other equipment such as an ADF or Omega.

• A 4096 code transponder with MODE C automatic altitude reporting equipment, except for helicopters operating at or below one thousand feet AGL under a Letter of Agreement. Note that this is also required inside the Mode C "veil" beyond the edge of the TCA.

What if you are operating perfectly legally with both a transponder and an encoding altimeter, you get half way through the TCA and your encoder conks out? Or you are outside the TCA just about to enter it when your encoder bites the dust? No problem...ATC is allowed when you notify them to authorize a deviation from the altitude reporting equipment requirement.

If you lose your transponder before you enter the TCA you are out of luck. ATC must have at least an hour's notice before they allow you to operate in that airspace.

Under both these scenarios the key word is **may**. ATC does not have to let you operate that way. It is up to them based on their workload, what kind of mood they are in and how nicely you ask.

As far as ratings go, you must have a private pilot rating or higher to land or take off from an airport within a Class B/TCA.

This brings up a few interesting points. If you are taking off and landing at South Fulton airport which is just southwest of Atlanta Hartsfield, can you do so as a student pilot? The airport lies well within the lateral boundaries of the TCA.

Sure you can. As long as you stay below 3,500 ft msl you aren't in the TCA, you are below the floor of it.

The second question is can a student pilot on a VFR cross country fly from South Fulton airport across the TCA to Stone Mountain Britt Memorial airport on the northeast side of town?

Of course he or she could! As long as they don't land or take off in the TCA, are properly equipped and follow ATC's clearances, they are perfectly legal to do just that. The rule wasn't made up to hassle student pilots; just keep them out of the traffic pattern at Hartsfield.

There are a couple of miscellaneous rules that may apply to you:

• If you are flying a large turbine engine powered airplane to or from the primary airport of a Class B/TCA you are supposed to remain above the floor of the TCA as long as you are within the lateral limits of the TCA.

A "large" turbine engine powered airplane would be by definition any airplane weighing more than 12,500 pounds.

This rule was made up primarily to protect the good old airliners. If the purpose of the whole TCA is to provide large, airline-type aircraft with added protection and separation from other aircraft it stands to reason that it would be a good idea to make sure that they stay in that airspace.

• You may not fly an aircraft in the airspace underneath the TCA at an indicated airspeed of more than 200 knots or 230 mph.

What is the speed limit within the TCA? 250 knots. This is quite different from the maximum speed limit in an airport traffic area of 200 kts or 1.3 times your clean stall speed. The reason for this is that the traffic mix in a TCA is mostly made up of turbine aircraft that regularly operate at 250 knots or more. Since all the traffic is capable of going fast and the airspace is crowded, why not let them go and come at a faster rate than you would at a smaller airport with a mix of jets, Cessna 150s and Aztecs?

Is there a minimum speed in a Class B/TCA? Not by regulation, but don't be surprised if the controller asks you to "maintain 170 knots until the marker." If you are operating at 80 knots on final at a busy place like O'Hare you can see the obvious problem if you're being followed by seven or eight jets whose minimum approach speed is 140 knots or higher. There's no need to worry if the controller has you zooming over the numbers at cruise speeds. There's probably 10,000 feet of runway in front of you to slow down in, and nobody said you had to make the first turnoff—though they'd like it if you did.

IFR Operations in a Class B/TCA

If you fly IFR using a properly equipped aircraft, into a TCA there is really nothing to worry about and no extra planning required. If anything, you are protected a little more in this crowded air because of all the rules governing it. From your point of view the operations are the same as what you do all the time—only perhaps more intense and pressured.

There are a few special techniques you can use while flying IFR in crowded airspace like a TCA but we will devote space to that later on.

VFR Operations in Class B/TCAs

As you have probably guessed by now or have discovered from experience, it is much harder to operate VFR in a Class B/TCA than it is to fly in the same airspace on an IFR clearance.

If you are flying VFR into a Class B/TCA, your first chore is to figure out exactly where the TCA begins. We went over this earlier in a general fashion but left out the practicalities. For example, if you are approaching the Atlanta Class B/TCA from the southeast from Milledgeville Georgia at an altitude of 6,500 ft MSL, how would you know exactly where the boundary of the TCA is? This is especially difficult if you don't have a DME (distance measuring equipment) on board, assuming you even knew the mileage figures.....notice that they are not shown on the chart.

According to the FAA: "Arriving flights must contact ATC on the appropriate frequency and in relation to geographical fixes shown on local charts. Although a pilot may be operating beneath the floor of the TCA on initial contact, communications with ATC should be established in relation to the points indicated for spacing and sequencing purposes." Over the past few years NOAA has started placing small pictures of recognizable landmarks on the TCA chart to help pilots not familiar with the area identify them. The Baltimore/Washington terminal area chart, for example, has drawings of the Washington Monument and Capitol building on it in the appropriate spots.

Many times, especially if you are flying into the primary airport of the TCA "sequencing" will mean that you are going to have to hold somewhere until ATC can find a spot to fit you into the inbound flow. It is a fact of life for the VFR pilot flying a slow aircraft that it is difficult for the controller to fit you into a traffic pattern that is flying at more than 170 kts.

Since you are VFR, ATC can't give you a holding fix like they can give IFR traffic. In other words, telling a VFR pilot to hold northeast of the Macy intersection, left turns, twenty mile legs won't work. They will usually give you some sort of easy to see visual fix like a shopping center or factory to hold VFR over until they can work you in. Or they'll simply tell you to remain clear.

When the controller gives you headings and altitudes to sequence you either into or out of a Class B/TCA they are doing so assuming that you will sing-out if they clear you into an illegal situation. In other words, it's your fault if you let them vector you into a cloud when you are flying VFR.

The problem with this, of course, is that it is nearly impossible to get a word in edgewise on the approach frequency. If you can't tell the controller about the cloud because of frequency congestion, what can you do?

You have no real choice here. You have to assume that good old "Pilot in Command" authority the government has bestowed upon you and do whatever is necessary to avoid the "emergency" of entering that cloud. Rest assured, when the controller sees you veer twenty degrees off the heading she just gave you, you *will* get a call on the radio from that controller. That's when you tell them about that cloud.

If you are leaving or transiting the Class B/TCA VFR the FAA encourages you to stay out of the airspace to the extent possible to you. They want you to either operate above or below the TCA's airspace or use one of the VFR corridors they have set up for that purpose.

That sounds pretty restrictive but remember if you do decide to get a clearance from ATC and fly your aircraft VFR directly through the TCA you're allowed...but with all the vectors necessary to get you through the airspace without hitting anybody it may take you an awfully long time to do it.

In the end, it's more expedient to get out of the Class B/TCA as fast as possible.

Secondary Airports in Class B/TCAs

Many times you will have no interest at all in flying into the primary airport of a Class B/TCA. As a matter of fact if you're smart you will probably avoid a primary airport in a large TCA unless it is impossible to do so. It is much easier in a general aviation aircraft to get into Peachtree-Dekalb airport in the Atlanta Class B/TCA than it is to fly into Hartsfield.

Peachtree-Dekalb caters to general aviation types and your speed will be compatible with the other traffic there—not to mention the fact that wake turbulence will be much less of a problem.

Unless you avoid the TCA airspace completely which isn't impossible but is difficult, you will have to deal with ATC to get in even if you are VFR only. It is no big problem. Just expect to get vectored around the traffic pattern of the primary airport in order to get to the airport you intend to land on.

Sometimes the vectors seem to take you well away from your intended place of landing. It will make more sense to you if you realize that many times the primary airport has traffic strung-out on final approaches up to twenty miles long.

Even if you are landing at a secondary airport, if you fly through Class B/TCA airspace you must have at least the equipment required for operating in that TCA and an ATC clearance of some sort through it.

Class B/Terminal Control Areas are a fact of life in the United States and will continue to be in existence for the foreseeable future. Dealing with them is no problem as long as you know the rules and follow them.

Next we'll look at the other features of airspace that exist at higher altitudes, but which are not discrete types of airspace.

Airways

So far, we've covered what you would find close to the surface in terms of controlled airspace. The surface is what most of us are concerned with, especially we pilots that fly small, light aircraft close to the ground and in busy areas.

As we climb to the higher altitudes we will find many more kinds of controlled and uncontrolled airspace for our perusal. The one GA pilots deal with most often is the Federal Airway, better known as the "Victor" airway.

In this case, the best way to explain the Federal Airways system in the United States would be to quote from the FAA's *Instrument Flying Handbook*:

"Each Federal Airway is based on a centerline that extends from one navigation aid or intersection to another navigation aid (or through several other navigation aids or intersections) specified for that airway. The infinite number of radials transmitted by the VOR permits 360 possible separate airway courses to or from the facility, one for each degree of azimuth. Thus, a given VOR located within approximately 100 miles of several other VORs may be used to establish a number of different airways."

Although most airways are made up of VOR radials, some are made specifically for operators with an RNAV (area navigation) capability.

Since, in order to be under control of Air Traffic Control the airways must be in controlled airspace, they all begin at least 1,200 ft AGL, the base of Class E/controlled airspace. The first layer of these Federal Airways are called "Victor Airways" and run from 1,200 ft AGL up to, but not including 18,000 ft MSL.

These "victor airways" are referred to by most airline and corporate jet pilots as the "low altitude structure." Although while operating on these federally mandated "highways of the sky" you are definitely in controlled airspace, during VFR conditions even the airline pilot or corporate jet jock is vulnerable to the scud-runner or VFR pilot who is lost.

In those parts of the western U.S. that have no "blanket" of Class E airspace covering the area, the airways run through their own corridors of controlled airspace.

Many IFR pilots spend most of their time on airways. For the VFR pilot, airways can be largely ignored—but there are two things you need to know about them. First is that the airway provides a convenient way to refer to your route when talking to ATC, and is probably fairly direct anyway. Second, and more important, is that the airways are where the IFR traffic is.

This is not universally true, of course. Increasing numbers of aircraft are equipped with area navigation, and as GPS becomes more prevalent the figure should skyrocket. An RNAV-equipped airplane operating IFR can turn up almost anywhere. Also, ATC may be issuing vectors to an IFR pilot, taking him off the airway.

Nonetheless, when on an airway a VFR pilot will probably be closer to more traffic than when off it. Also, it pays to be aware of how airways cluster together at VORs. If the airways are like highways, the VORs are like intersections, with lots of traffic directly overhead.

Jet Routes

As we discussed earlier on the advent of the jet airliner caused ATC many added problems in terms of separating jet aircraft from the slower prop-driven aircraft and improving the survivability of both slow and fast classes of aircraft in the same skies.

At 14,500 feet MSL we find the beginning of the Continental Control Area. The Continental Control Area extends from 14,500 MSL upwards through infinity. It is conceivable that the Moon is in the Continental Control Area if you read the FAA's definition literally because it gives no upper limit. Under ICAO, the CCA will become just one more variant of Class E airspace that has its floor at 14,500 feet.

The Continental Control Area is also significant because it is the top of all uncontrolled airspace in the United States. If you're above fourteen-five you are definitely in controlled airspace of some kind.

In terms of required equipment on your aircraft the Continental Control Airspace area is two thousand feet above the lowest altitude you can operate an aircraft in the United States without a 4096 Mode C Transponder. At 12,500 feet, whether in controlled airspace or not, you must be transponder equipped and it must be operating.

Jet aircraft usually begin their cruise segments at or above 18,000 ft MSL where the advantage of having jet engines comes to the fore. Much below 18,000 jets are gas-hogs and burn more fuel than a comparable turbo-prop or piston aircraft. When jets do cruise though, they really cruise. It is not unusual at all for your average, run of the mill airliner like the Boeing 727 to fly in cruise at 560 knots or about 0.86 the speed of sound.

Above 18,000 feet MSL all pilots set their barometric altimeters to 29.92 inches of mercury or 1013.2 millibars and enter the realm of the "high altitude structure."

Above 18,000 feet MSL the altitudes are no longer named after how high they are above sea level, they are called "Flight Levels." In addition to making the altitude easier to understand when spoken over the radio they are called flight levels for the simple reason that unless the altimeter setting on the ground you are flying over is exactly 29.92 inches you have no idea of how high you really are above sea level.

Don't discount the fact that altitudes are easier to understand when given in flight levels. "Delta 123 climb to and maintain flight level three-one-zero" sounds a lot clearer than "Delta 123 climb and maintain thirty one thousand feet."

Since the members of this high altitude "club" travel pretty fast over many different areas with differing altimeter settings it is considered safer for traffic altitude separation to have everyone set their altimeters to the "standard" setting. On some days, when the altimeter setting at the lower levels is below 29.92 inches, the lower Flight Levels are unusable.

Class A/Positive Control Area

Up in the higher altitudes lies Class A airspace, which is much simpler simply because it's so restricted. Most of the rules pilots must remember when it comes to lower airspace really apply to VFR pilots—in a sense, flying IFR is much simpler than flying VFR.

Class A Airspace Summary		
Basic VFR Minimums		
Not applicable.		
Was Called	ATC Clearance Required	Equipment Required
Positive Control Area (PCA)	IFR ops only.	Appropriate for IFR, plus DME
Comm Required	Separation	Services
Yes, with ATC	All	Positive ontrol of all airspace.

All operations in Class A airspace are conducted IFR. Therefore, none of the VFR rules and exceptions that the lower levels are riddled with apply. All the altitudes between 18,000 feet MSL and Flight Level 600 are in Class A/Positive Control Airspace.

Positive Control means exactly what it says. If you want to operate an aircraft legally in the airspace above FL 180 you must be instrument rated and qualified, be flying a suitably equipped aircraft and be operating on an ATC IFR clearance. The logic of this rule is pretty simple. At the higher altitudes aircraft routinely cruise at speeds of over 600 miles per hour giving them a potential closure speed of over 1,200 miles per hour. There must be some mandatory protection at these speeds to lessen the risk of mid-airs.

High altitude or Jet airways are fewer and further between but they do have the advantage of longer range reception for your nav radios, they are much less crowded than the low routes and you are under the watchful eye of a controller who's job it is to insure your separation from all that nasty traffic out there.

The jet route structure runs from the base of Positive Control Airspace up to flight level 450. Above that altitude you find mostly military traffic although with the advent of the higher flying business jets and airliners like the Concorde civilians are poking their noses closer and closer to outer space.

In the lower flight levels there is no more equipment required than the transponder you need to fly above 12,500 and nav radios capable of keeping you on the airway. Above Flight Level 240 you must also have a usable DME (distance measuring equipment) receiver. Although many airliners and military aircraft are equipped with much, much more navigational equipment than listed by the FARs you might be surprised to learn that very many of them are equipped with the bare minimum required for flight at these altitudes. The standard, stock Boeing 727 used by just about every major airline in the United States is usually equipped with two nav-coms, two DMEs, a transponder and one ADF! They are reliable and they do the job. Most people that fly aircraft like Barons might be interested to learn that the DMEs on the 727s don't usually even have a ground speed read-out. We still use the "time the mileage and divide by six" method to determine our approximate ground speed.

Special Use Airspace

Even though this book is dedicated to your survival in congested airspace we should discuss the different forms of "special use airspace" you will find and have to deal with in your travels.

Isn't special use airspace just something you run into out in the desert, like bombing ranges? Yes, you do run into areas like that out in the boonies but you also run into them in crowded areas like TCAs. Would you believe there are numerous restricted and even prohibited areas in the Washington TCA? Yep, and you have to try really hard to avoid them too. There is a prohibited area (the White House) right off the departure end of runway 36 at Washington National.

These are not discrete classes of airspace like those we've been discussing. Basically, they're places where the rules may be altered from time to time for special purposes. If you want to play it really safe, just avoid them altogether. That's neither expedient nor practical in many areas, however, so it pays to understand how these areas work.

Let's run a quick review of the kinds of special use airspace and what you must do to deal with them.

• **Prohibited Area:** These are pieces of airspace clearly depicted on both IFR and VFR charts. You are absolutely **not allowed to fly into or through these areas**. This is important to remember because there are more than a few of these that will get you hurt if you trespass.

There are very few prohibited areas but they do exist. Usually they encompass places like the Capitol building or key military locations. Do everything you can to avoid flying through these areas.

• **Restricted Areas:** These are areas where you are prohibited from flying only part of the time. Gunnery and bombing ranges as well as certain military airspace make up restricted areas. If you have any questions at all about whether or not you can fly through one of these areas ask ATC.

Restricted areas are also depicted on aeronautical charts. Along with their dimensions are there hours of operation and what controlling facility is in charge of it.

Even though it seems the only times these areas are "Hot" (active) is when you need to fly through them to avoid a line of thunderstorms, keep in mind that to fly through one when it is in use can be hazardous to your health. They are very often using real bullets and missiles in these places and the military guys aren't looking for you...they're busy. If you have any doubt at all about restricted airspace, stay out of it.

• **Warning Areas:** These are areas that are really restricted areas but can't be called that by the United States government because they are officially out of its jurisdiction. Warning areas are pieces of airspace that lie outside of the legal three mile limit offshore. Since the airspace is outside of the United States and is over international waters it cannot legally be named a restricted area.

In your mind it should be considered the same thing because the same kind of activity goes on in a Warning Area that transpires in a Restricted Area. We're talking supersonic fighters shooting real missiles and guns. Stay away from them.

• **Military Operations Areas (MOA):** These are blocks of airspace used by the military primarily for training flights.

If you are IFR you can be cleared through this airspace by ATC as long as they are sure they can give you separation from the participating military traffic. For example, if the military guys were going to do all their flying below flight level 240, ATC could clear you through at flight level 330 without a problem. If there is any problem ATC is required to route you around the MOA.

If you are flying VFR you are required to avoid an MOA if it is active. The FAA recommends that you call ATC anytime you are within 100nm of a MOA to see if it is active.

Military pilots operating in a MOA are usually performing aerobatics because that is the main reason this airspace was created. In a MOA the pilots participating are exempt from FAR part 91.71. That is the regulation that prohibits aerobatic flight on Federal Airways. These guys aren't looking for you...it is your job to look for them.

• **Alert Area:** These are parcels of airspace depicted on charts to inform you about unusual activity like flight training or parachute jumping that is going on. It is the responsibility of both you and the people doing this unusual activity to look out for each other.

There you have it—the nation's airspace in a nutshell. This is the playing field on which all pilots travel, and it's wise to never ignore it. Keep track of where you are, and you'll not only know what's required of you but what services are available to help you.

Next we'll get into the nuts and bolts of how to survive in the airspace structure and how to make use of its features to your best advantage.

Communications Practices

The real heart of ATC is not the procedures, rules, or structure; it's *communication*. The controller simply has to be able to talk to the airplanes he's trying to keep separated to make the system work effectively.

Outside of Class A, B, C and D airspace a pilot is not required to be in radio contact, and of course the system is designed to allow for traffic that isn't in contact with ATC. But when things get crowded, the system can have problems.

The best example is the 1986 Cerritos, California mid-air collision, in which a DC-9 and an Archer ran into one another. The Archer pilot wasn't in contact with ATC at the time; had he been, he would have been vectored away from the jet. Of course, the DC-9 pilot was supposed to have been kept clear of the VFR traffic, but with both pilots hooked into the system the chances of conflict would have been greatly reduced.

The Pilot's Perspective

From the pilot's point of view, the rapid-fire communications used in congested areas are the single most confusing and frustrating aspect of operating in crowded airspace.

Novices often find that the controllers are talking so fast that it's hard to follow what's being said. Even experienced pilots find that it's virtually impossible at times to get a word in edgewise.

Nothing is more important to a pilot flying in congested airspace than good communication. Keeping things clear and professional will make getting through a crowded area easier, and will keep the controller happy (it's not in a pilot's best interests to have the controller get impatient with him).

Talking on the radio is an acquired skill, and ordinary training will teach the basics. But, there are techniques that, if practiced regularly, will improve a pilot's communication skills.

Keep Current

When a pilot doesn't fly, his skills deteriorate. It's necessary to keep current in order to fly an airplane safely, and the more the better: FAA's legal currency requirements are woefully inadequate in terms of maintaining flying skill.

The same is true of communication skills. One of the first things to go when you get rusty is your radio technique. It's important to be aware of this, and to be prepared to listen more closely and to pay more attention to your transmissions if you haven't flown in a while. If you fly a lot in high-density areas and talk to ATC regularly, you'll have an easier time of it than if you only occasionally ask for flight following and rarely enter areas where you must talk to approach controllers.

It's easy to keep sharp on the radio: simply use it as often as possible, even when it's not required. Even when ATC can't accommodate a request for traffic advisories, listening in is a good idea: it can keep you alert to traffic in your area.

For VFR pilots, the benefits are even greater. Much of instrument flying is nothing more than proper communication. Learning to use the radio like a pro will go a long way towards preparing the pilot for an instrument rating. Not to mention that paying attention to communications helps keep a pilot sharp: motoring along at cruise leaves a pilot with relatively little to do, and using the radio will keep his mind focused.

Basics

It's surprising how many pilots simply don't use the radio correctly. On any given day, you'll hear pilots blithely stepping on transmissions, undermodulating, shouting, or keying the mic improperly.

The fundamentals are simple, but are worth repeating here. First, it's necessary to position the mic properly. Most pilots these days use headsets, but the old rule still applies. The mic must be close to your mouth for you to be heard. A good rule of thumb is that if you can't kiss the microphone, it's too far from your lips.

For those who use handheld microphones, the same thing applies, with one addition: noise-canceling mics have a small hole on the back of the microphone that must be left uncovered for the mic's noise canceling to work properly.

A small foam windscreen is a worthwhile addition to a headset's microphone. It improves the sound quality remarkably.

Next, keep your voice down. Just because the inside of your cockpit sounds like a boiler factory doesn't mean you have to shout into the microphone to be heard. If your passenger can hear you talking on the radio, it's a safe bet you're talking too loudly (or that your engine has just failed).

Many pilots start talking at the same time they press the push-to-talk (PTT) switch, particularly when they're trying to do things fast. This is a mistake, because it often results in the first word of the transmission being cut off. Instead, push the button first, then speak. There's a lot of pressure in crowded areas to keep the time spent transmitting to an absolute minimum, but the extra tenth of a second it takes to do it right won't make much difference to your time spent on the air.

Lastly, listen before speaking. This is particularly important in a crowded environment, where a controller may be talking to a half-dozen other pilots. It's necessary to listen and wait for the right moment to jump in, or you'll step on someone. Often, it can't be avoided—remember, there are probably others on the frequency waiting for the same opportunity you are, and they're likely to call the controller at the same time you do.

Conversely, you can't be too timid, or you'll wait forever. There will be a kind of rhythm to the flow of radio communication, and it's possible to pick up on it and make the call at just the right moment to fit into the flow.

Listening

There's no way a pilot can follow all of what's being said on a busy frequency and still fly the airplane. Even listening intently for your own call sign can be tiring and distracting after a while.

It's not necessary to try to listen to everything a controller says in order to catch the transmission intended for you. The first word out of his mouth will be the type of airplane; if it's what you're flying, then listen for the call sign. All other transmissions can be ignored completely. It can be thought of as a trigger: as soon as you hear your make of airplane, start listening. In practice, it becomes almost automatic.

The next problem has to do with gleaning the important information from a radio transmission. Controllers often issue relatively complex directives very quickly, and you're expected to get it right and read it back the first time, every time.

There's a simple trick to deciphering even the most rapid and confusing of radio transmissions. The key is that they're predictable.

Controllers will always say the same sorts of things in the same order. In other words, the form of the transmission is always the same, only the content varies.

In addition, the kinds of transmission a controller will be making will vary according to what kind of controller he is. For example, you will never hear an enroute controller giving an approach clearance, nor will a tower controller issue an altitude or heading change. What this means for the pilot is that he can expect only certain types of information to be transmitted to him.

In a sense, then, the pilot knows what the controller is going to say even before it's said. That anticipation allows the pilot to concentrate on the important part of the transmission—the content—without having to pay close attention to every word.

Let's put this in context: we don't mean that the pilot should block out what the controller is saying. It is possible, however, to pick out the parts of a routine transmission that are actually important and need to be acted on. This will increase comprehension and reduce the pilot's workload.

The following is a good example of a transmission that's long and complex. It's easy for an unprepared pilot to miss something important in it because of that.

"Cessna 1234, two from the outer marker, turn right heading 140, descend and maintain 2,000 until established on the final approach course, cleared for the ILS 16 approach, report outer marker inbound."

That transmission contains three separate instructions, a clearance, and a position for the pilot to digest. Not only that, but the pilot won't know exactly when the transmission is coming, and so might be a bit behind when the controller fires it off.

Passively listening and waiting for the clearance is not a good way to pick up the entire meaning of the transmission. There's so much there that an unprepared pilot can easily get overwhelmed and miss the whole thing. However, a pilot who anticipates it will have no trouble.

As with other transmissions, the form of an approach clearance is always the same. The pilot already knows that he's being vectored for the ILS 16 approach, and if he's doing his job right he knows that he's getting close to the outer marker. So he knows that the controller is about to issue the clearance, and he'll probably be ready to note the two really important things about the clearance—the heading and altitude. Our favorite technique for this is to use the ADF card to note the altitude (e.g. set it to 200 degrees for 2,000 feet) and the heading bug on the directional gyro for heading assignments. If the airplane doesn't have a heading bug, we jot the course down.

So what about that transmission is really important to the pilot? Basically three things, the heading to steer, the altitude to descend to, and the instruction to report the marker inbound. The rest can be essentially ignored. That makes getting the gist of the transmission much easier.

Once you've got it, you can read it back:

"Cessna 1234 is two from the marker, right to 140, descend and maintain 2,000 'til established, we're cleared for the approach and will report marker inbound."

Again, most of this is just form. The pilot just says the standard phrase and inserts the appropriate numbers.

Why can the pilot filter out the rest of the transmission? Mostly because it's either part of the form or information the pilot already has. As noted above, the pilot already knows which approach this is, and if he wasn't cleared for the approach the controller wouldn't have turned him towards the airport and told him to descend—the transmission would have sounded entirely different.

The fact that he's two miles from the outer marker is important only in that it alerts the pilot that he's close to the marker; the number of miles isn't really important. Besides, he probably already knows that if the airplane is equipped with DME.

The phrases "until established on the final approach course" and "marker inbound" are part of the standard phraseology.

So, what's left? A heading, an altitude, and the requirement to report the marker. That's much simpler than trying to follow the entire transmission, particularly if the workload is already high.

Naturally, the pilot doesn't think all this through—that would take too long. Essentially it's a technique for picking out only the important words in the transmission as it's happening.

The same technique can be applied to any transmission. For example, when calling a control tower, you know ahead of time that the controller is going to tell you which leg of the pattern to enter on, which runway is in use, and which direction the pattern is going. If the field has no ATIS, there will also be wind information. If you've been listening, you probably know most of this already, so you only need to listen for what the controller wants you to do.

Another example is a typical traffic advisory. It will sound something like this:

"Cessna 1234, traffic 10 o'clock and four miles southwest bound, a Beech 1900 at 6,000."

The moment you hear the word "traffic" you know you'll be looking for something outside the cockpit. The very next words, "10 o'clock," tell you which way to look. Often, by the time the controller has finished the sentence, you've already seen the traffic. This is because you already know what he's going to say, and so don't have to think about it: all you need to listen for is the information contained in the sentence.

Another thing about anticipating what a controller will say is that the pilot can usually tell what sort of transmission it is within the first few words.

The traffic report is a good example, as is the ILS clearance discussed above. In that case, as soon as the controller said "two from the outer marker" the pilot knew it was a clearance and was listening for the heading and altitude.

This is the same principle as picking out the transmission intended for you from the general babble by listening for your airplane make; these are examples of more trigger words.

Speaking

During training, pilots are taught to structure their transmissions in a certain way to get the point across to the person on the other end of the line.

Basically, it goes like this: who you're talking to, who you are, where you are, and what you want. For example, a call to an airport control tower might go like this:

"Bridgeport tower, Cessna 1234 is 15 miles northwest of the field with ATIS information Zulu, inbound for touch and goes."

That's fine, and works well in that situation. But using the standard structure right off the bat when talking to a controller in a busy Class B area like Los Angeles isn't such a good idea:

"Los Angeles approach, Cessna Skyhawk 1234 is over Ventura VOR at 6,500 feet, requesting transition of the Los Angeles Class B area enroute to Ciraco Summit."

That's quite a mouthful on a first transmission, and is likely to get a terse "remain clear" instruction. Even worse is forcing the controller to play "20 Questions" by not including the information he needs at the appropriate time:

"New York Approach, Cessna 1234."

"Cessna 1234, go ahead."

"New York, Cessna 1234 is over Stamford, requesting clearance into the TCA."

"Cessna 1234, say your destination."

"Cessna 1234 is enroute to Frederick, Maryland."

And so it goes. That whole process takes far too long, and is a real waste of everyone's time.

Far better and more appropriate in the context of busy airspace is to use the first transmission to simply let the controller know you're there and that you want something:

"Miami Approach, Cessna 1234 with a request."

This has several advantages. First and foremost, it gets you off the radio quickly. The controller must give priority to the airplanes he's already handling, and you're essentially asking to break in and give him more work to do. Spewing forth your entire life story on the first transmission is neither necessary nor appropriate. Another advantage is that by saying you have a request you alert the controller to the fact that you're a new customer.

When the controller has time, he'll say:

"Cessna 1234, go ahead with your request."

Now is the time to give the entire picture. The controller is listening to you, and the other pilots on the frequency know you're about to make a relatively long transmission and will be less likely to inadvertently break in. Be sure to give the controller everything he needs on the first transmission. That does *not* include your entire route: the controller wants to know where you are now, what you want of him, and your destination so he can vector you appropriately.

The controller will tell you one of two things. Either he'll tell you to remain clear of the area, or he'll give you a transponder code. Therefore, be prepared to dial the code in, or at least write it down and read it back. Remember, anticipating what he'll say is the key to picking it out of the chatter. You know he's not going to give you a vector or altitude until he has you in radar contact, and that can't happen until you are squawking a code. So, you can forget about heading or altitude changes, and concentrate on listening for a transponder code.

In addition to using proper phraseology, it's very important to decide on your exact words *before* you key the push-to-talk switch. This is critical on a crowded frequency. There's no place there for "ums," "ers" and dead air. Think about what you're going to say (not just what you'll be asking for, but the *exact* words you'll be using), then say it. This becomes second nature after a little while, but it's one of those skills that degrades rapidly if not used.

Catch-22

A VFR pilot who is concerned about handling himself on the radio is faced with a dilemma: how to get used to it without diving right in and suffering the trial by fire.

An excellent way to prepare for operating in an area with heavy radio traffic is to listen to some. There are several ways to do this. Several good tape programs are available from pilot shops and aviation publishers. These offer advice on radio technique along with recordings of ATC communications.

Or, you can make your own. Many portable intercoms can be plugged into tape recorders, and will pick up whatever is coming over the audio circuit. Just tune in the local approach control frequency and listen. Reviewing the tape in your car will teach you a lot about how communication works.

5

Organizing Yourself

F lying into and through congested airspace can be made much easier if you're on top of the situation, and that requires organization. Fumbling with cantankerous charts in heavy turbulence while trying to find a frequency is a pain in the neck at best.

In this chapter we'll look at various ways to keep your ducks in a row in the cockpit, from specific methods to general advice on how to keep the information monster under control. We'll cover the various kinds of accessories you might want to use in the cockpit, and look at some specific examples of organization schemes that have worked for us.

First, though, we'll cover the foundation of being an organized pilot, which is the ability to use what's available to you in the cockpit to your best advantage.

Cockpit Resource Management

As a pilot, you have a task to perform, namely to conduct a flight safely through crowded airspace. To accomplish this, you have a number of tools at your disposal, including your piloting and navigational skills, your flight and nav instruments, your information sources (charts, books, Flight Service and the like) and the services of air traffic control. In addition, you may have a second pilot on board to whom some of your duties can be delegated. All of these are resources to be drawn upon in order to complete the flight. Doing so effectively is known as *cockpit resource management*.

The idea of CRM popped up in the '70s as a result of NASA research into air carrier accidents involving "pilot error." The studies found that professional pilots have very high levels of technical skill, but that in many cases the crews were unable to handle emergencies effectively.

There were seven specific problem areas found: preoccupation with mechanical problems; inadequate leadership; inadequate monitoring of the flight; failure to delegate tasks to others; failure to utilize all available information; failure to communicate intent and plans; and failure to set priorities.

As a result of the studies, the NASA human factors researchers devised five general rules to be followed in the cockpit:

• In abnormal situations, the first order of business should be to decide who flies the aircraft and who monitors or works on the problem.

• Positive delegation of monitoring duties is as important as positive delegation of flying duties.

• The pilot flying must not attempt to accomplish secondary tasks during busy portions of a flight.

• Whenever uncertainty or conflicting opinions of fact occur, such as a misunderstood radio transmission, the conflict must be resolved unequivocally using external sources of information. (In plain English, that means asking for a repeat or clarification.)

• If any crew member doubts a clearance, procedure or situation, he or she is obligated to make that doubt known to other crew members.

Obviously, these guidelines are slanted towards air crews, not single-pilot operations. Nontheless, there's some very good advice here for any pilot, no matter what he or she flies.

For example, one of the rules says to not attempt to perform secondary tasks during the busy portions of a flight. If you're flying into a major terminal area surrounded by Class B airspace, you need to be paying attention to ATC, not fiddling with a sectional chart. It's of the utmost importance to remain aware of your priorities and not let the situation get away from you. The object of organizing your cockpit is to make your primary task easier; keeping everything in order inside the cockpit will let you keep your eyes outside and your thoughts on the task at hand.

Delegating tasks can work even without a second crew member.

If you have a passenger, get him or her involved. You don't need to be a pilot to look for traffic or stow charts.

A pilot flying in a congested area must stay ahead of things and maintain situational awareness. Where are you? What, exactly, are you doing now and what will you be doing next? (One of our favorite sayings is that the two most important things during a flight are the next two things.) If you begin to lose track of what's going on and get behind, admit it—it's far better to 'fess up to the controller and get out of there than to go fumbling your way through a hectic terminal area without a clear idea of what you're doing.

Develop a Method

The most important thing to realize about organizing yourself as a pilot is that there's no single "correct" way to do it. Every organized pilot has his or her own pet cockpit organization scheme and method for gathering preflight information. There are many mixes of chart systems, supplemental information sources, checklists, mnemonics and gadgetry to choose from in cooking up your own recipe for staying organized. It can be as complicated or as simple as you like: some pilots like to fly with just a notepad and chart, others like to have compartmentalized flight logs with a place for every conceivable bit of information they'll need. Some like to use checklists exclusively, others rely heavily on tried-and-true mnemonic tricks. With today's sophisticated loran and GPS units, it's even possible to fly VFR in a "paperless" cockpit—though we wouldn't recommend it.

Flight schools tend to teach a single, rigid method, usually based entirely on government publications. (Remember showing your laboriously completed flight log and marked-up chart to your instructor before a cross country?) There's certainly nothing wrong with that, but many pilots find that other methods may work better for them.

The first step in getting organized is to develop your own method, then stick with it. Experiment at first with various kinds of flight logs, checklists and charts until you find a setup that you're comfortable with. After that, you should use your system every time you fly; the idea is to make using an organization scheme a habit, something you do as a matter of course.

Regardless of the exact scheme you devise, it should have certain characteristics:

• **Simplicity.** It's tempting, at first, to overdo things. It's possible to spend so much of your time refolding charts, keeping logs up do date, doing new time and fuel estimates and cross-checking positions that you fail to keep up your visual scan for traffic. That's potentially fatal in crowded airspace. By the same token, it's possible to spend all your time keeping the airplane on course and watching out for the other guy at the expense of staying organized. If you completely ignore organization, you could get caught with your pants down just when you need a critical bit of information at your fingertips. Any organization scheme should fit into your priorities as a pilot and serve you, rather than the other way around.

• **Ease of Use.** Your scheme should allow you to come up with needed information without having to hunt for it. You shouldn't be busy digging in your flight case looking for a terminal chart when you're about to penetrate a Class B area—it should already be at hand. Everything you're likely to need for a flight should be accessible before you even start the engine. Any organization scheme should take that into account.

• **Completeness.** Make room in your method for everything you need. The bare minimum (i.e. a current chart and a pencil) may get you where you want to go, but to be efficient and safe you should avail yourself of far more than that. The goal of any organization scheme should be to reduce your workload in the cockpit so that you can focus on the business of flying. Anything that helps to achieve this goal is worthwhile, and should be used. Anything else is extraneous, and you should question whether you really need it or not.

• **Flexibility.** You should be able to deal with unexpected changes to your plans. A meticulously detailed IFR flight log filled out in advance is of little use if ATC decides not to let you fly the route you had planned on.

• **Portability.** Again, it's tempting to go overboard. There's no need to carry a duffel bag full of charts, books, plotters and calculators with you all the time. Many professional pilots carry precious few charts and approach plates, but they're selected with great care.

• **Security.** You could just pile everything on your lap, but as soon as you hit some turbulence it would all go falling to the floor. You could also strap everything down so tight it would be impossible to get at any of your charts. Most pilots come up with a compromise of some sort, and thought should be given to this.

Next we'll look at some general guidelines that apply no matter what the details of your personal organization scheme.

Plan Ahead

If the goal of organizing yourself is to reduce the workload in the cockpit, then it makes sense to do as much of your work outside of the cockpit as you can. It's possible to get a surprising amount of your work done before you leave home.

It's a good idea to give yourself a preflight briefing that goes beyond just checking the weather: examine the route to see who and what you'll be dealing with along the way. Make plans for the most effective way to complete the flight. Figure out what to do if things don't go your way.

First and foremost, you should get hold of as much information as you possibly can before going anywhere near that busy terminal area—or, for that matter, near the airplane.

Before making a flight that you know will take you into crowded airspace, gather all your charts and books and go over them. Don't be lulled by recent experience—sure, you may have flown through the same area last month, but what if FAA has changed the boundaries of the airspace? You just might blunder into an area where you're not supposed to be without realizing it, or be trying to call a controller on a frequency that's no longer used. Getting into the habit of examining the charts also keeps you from relying too much on memory.

Also go to other printed sources, such as the Airport/Facility Directory. This is a handy source for frequencies and any important information about your destination and points you'll be flying over. There are several commercial airport directories that also offer tips on VFR procedures into and out of crowded airports that aren't found in government publications (the JeppGuide [formerly TannGuide] series is one).

It helps to pick pertinent information up and record it someplace that's easy to find—for example, the frequencies for the departure and arrival airports, along with enroute approach controls should be written down someplace handy so they don't need to be searched for. This information is on your charts and in other books, but that's not the best way to get at it in the cockpit. AOPA's comprehensive volume of airport information *Airports USA* may be a good reference for airport and business information, but it's also the size of a major metropolitan telephone book. Why carry it along when you can just jot down what you need to know?

On the subject of airport information, finding out where the FBO is on an unfamiliar airport can be useful. At a large airport, even turning the wrong way off the runway can cause you real headaches trying to taxi over to where the FBO is. Many commercial airport guides show this on an airport diagram, and give phone numbers and service information. A call to the FBO before the trip will verify that the information is correct, as well as giving you the lowdown on parking, landing fees and ground transportation. (There have been many accidents in which pilots assumed the FBO would be open or that they'd have fuel available. When the pilot landed and found that not to be the case, he'd try to make it to the next airport down the line and run dry on the way. A simple phone call can keep that kind of thing from happening.)

The route, itself, can be marked on the chart to make it easy to find in flight. We're in the habit of marking all the pertinent VORs and frequency boxes on our Jepp enroute with yellow highlighter—that way, when it's time to change nav frequencies, we don't need to search for them on a complex chart.

If you're an IFR pilot or VFR pilot who uses approach plates, pull the pertinent ones and put them on your clipboard (Jepp) or mark the place in the book with a rubber band (NOS). This keeps you from having to fumble around in the book looking for the right airport's plates.

Get at least one weather briefing, and record the information in a legible manner on a separate piece of paper. Better still, get two briefings: the first several hours before your trip, the second when you're ready to file your flight plan. The first briefing gives you a chance to get the big picture and lets you record any notams. It also gives you the winds aloft forecasts, so that you can make accurate time enroute estimates. The second briefing nails that information down, and lets you concentrate on the details.

Don't ignore "non-aviation" sources of information. One of our favorites is *The Weather Channel,* part of basic cable TV service in most parts of the country. It has radar maps that are at least as good as any you'll find in a flight service station and is a good way to familiarize yourself with the weather situation before you call Flight Service. There's also the venerable *AM Weather,* found on many Public Broadcasting Service TV stations early in the morning. This program is specifically oriented towards pilots.

Another excellent means of getting advance information is through DUAT or one of the commercial computerized weather providers. Using your computer can save you time and hassle, and you can even file your flight plan. One of its primary advantages is also one of its big disadvantages: lack of interpretation. Live briefers can help you with some feeling of how the weather picture will affect you: a computer printout can't. A computer printout also can't influence you subconsciously the way a briefer might.

Lastly, there are many commercial fax services. These can provide you with a staggering variety of official NOAA weather charts exactly like those pinned to the bulletin board in Flight Service Stations—provided you have access to a fax machine. We've found these services to be very useful, since the charts they provide are much easier to interpret than a verbal or text-based weather briefing.

At the end of the process you should have all the information you need recorded in a form that you can use in flight (this includes pertinent weather information, frequency and airport information, a completed flight plan, and possibly a flight log) and all the charts and plates you'll need ready to go.

All of this advance preparation serves two purposes. First, it familiarizes you with the area you'll be flying into before you leave. Second, it keeps you from having to fool around in the cockpit trying to get things organized and fly the airplane at the same time. The more you do ahead of time, the less you'll have to do in the air—and the more time you'll be able to spend flying the airplane, looking for traffic, and dealing with ATC.

In the Cockpit

Once on board, you should first take a moment to set everything up. The more you do before engine start, the less costly the flight will be. (In the winter, of course, you may want to start the engine first to let it warm up while you get yourself ready.)

On your lap or within easy reach should be your list of frequencies and other information, your weather information, whatever charts and plates you'll need, someplace to write fresh information (some pilots like to use the chart) at least two pens or pencils, and the aircraft checklists—plus your own, if you've devised any. The charts should be refolded to show at least the first part of your route. Sectionals and enroute charts are larger than the average cockpit, and they can be a nightmare on a bumpy day if you have to wrestle with them in flight. Any supplemental information, such as airport guides or charts you won't be using on this leg, should go in the back seat—keep anything you don't need immediately out of your way.

This is a good place to get your right-seat passenger involved, if he or she is interested. The passenger can be a valuable resource for you, even if they're untrained. They can watch for traffic, keep an eye on engine instruments, and handle charts for you. Keeping them busy is entertaining, and even helps fend off airsickness somewhat for those that are susceptible.

In Flight

If your organization scheme is well designed, it should work for any flight you make. You should use it as a matter of course, always recording the same information in the same places and keeping things organized in the same way. By doing this you'll never need to scramble about looking for a frequency or trying to find a needed piece of paper.

On a given cross-country, you may well pass off the area of one chart onto another. Since the first chart isn't being used anymore, there's no need to keep it handy. Putting it away will keep the cockpit neater and there will be one less thing to worry about. This also goes for approach plates having to do with the airport you've left.

The most hectic time for flights into congested airspace tends to be the last few minutes approaching the destination airport. That's when you need to be the most alert, and when you'll have the most to do in flying the airplane. On a typical IFR flight, by the time we're established on the approach, the only thing we have in front of us is the approach plate—everything else is stowed, but easily reachable. An approach is no time to be taking notes, and any enroute charts are no longer needed. For VFR pilots, you can pretty much stow everything once the pattern is in sight. By then you should have dialed in both tower and ground frequencies, and have no further need of notes or charts. This clears the decks and lets you look for traffic and concentrate on flying.

As noted above, taking notes and keeping records should not take an inordinate amount of time in flight. Remember, you're trying to get through a crowded area, not keep a picture-perfect log. In fact, many (if not most) IFR pilots we know don't even bother with actual flight logs. Students have these drilled into them from the moment they start flying cross-country, and they can be very useful for keeping track of long VFR legs—but in an intense radar-controlled terminal environment they're not really worth the effort.

Components

The individual parts of your organization scheme will consist of charts and other information sources, paperwork aids such as flight logs, and accessories such as kneeboards, flashlights, computers and the like.

We encourage you to try out a variety of products to see which meets your needs best. You can find everything you could ever need in the cockpit on the shelves of your local pilot shop or in a mail-order catalog. Alternatively, you can devise your own forms and logs. Rather than outline a single scheme, we'll run through the various options available. You can pick and choose as suits your own style.

Don't Ignore the Airplane

Several of the more useful items for keeping organized in the cockpit are right there in front of you every time you fly.

Staying organized in the cockpit really means managing information. There are radio frequencies to keep track of, assigned headings, altitudes to climb or descend to, times to switch tanks, and so forth. Many of the instruments found in the typical cockpit can be used as "teasers" to keep this information in front of you without you having to record it.

For example, the ADF card can be used as an altitude reminder. Just set the assigned altitude (or decision height, or MDA, or whatever) to the top of the compass card. If the MDA on your approach is 950 feet, set the ADF compass card to 95 degrees. This is particularly handy on an approach, when the controller issues you a final altitude assignment.

Many heading indicators are equipped with heading bugs. Whenever you get a vector from a controller, the bug should be set and followed.

Modern "flip-flop" radios let you set two frequencies, one active and one "on deck," or standby frequency. Rather than immediately writing a frequency down, just dial it into the radio as soon as you've got it. For many of the more hectic parts of the flight (departure and arrival) you can stay one step ahead when you know the frequency in advance.

Various nooks and crannies might exist that are perfect for holding certain critical items. Your charts might fit very well between the edge of the glareshield and the windshield, for example. Don't assume you must keep everything on a clipboard or kneeboard. Be opportunistic and experiment.

Kneeboards and Clipboards

You'll be writing quite a bit in the cockpit, so it's necessary to have a surface to write on.

The pilot catalogs are chock-full of specialized kneeboards and clipboards, some of which are ludicrously expensive—$60 or more. Even the cheapest, which are really little more than glorified clipboards, run $15 and up.

Frankly, there's no need to spend even that much, unless you want some special feature or other. A dime-store variety clipboard does the job just as well. If you want to get fancy, you can glue a couple of strips of foam rubber to the back to keep it from slipping off your lap. If you simply must have a kneeboard, a minimal amount of ingenuity, a hacksaw blade and some elastic can produce one that's just as functional as anything sold in a pilot catalog for a small fraction of the cost.

Our personal preference is for an aluminum clipboard/"forms holder," sold in office supply stores. This is a shallow aluminum box with a lid that serves as a clipboard. Inside the box we keep charts we're not using at the moment, spare pencils, a penlight, and weather information. The only drawback is that you have to remove it from your lap to open the lid, since the yoke gets in the way.

There are also variants on the kneeboard that clip to the yoke. We've found these to be awkward at best—the board makes the yoke top-heavy, it gets in the way of the instruments and it puts everything too close to your chest.

Flight Bags

Again, it's possible to go out and buy "pilot luggage" that costs way too much. Is a special ballistic nylon bag with dedicated compartments for your headset and handheld radio really worth eighty dollars? We think not.

Our personal flight bag is a soft-sided canvas attache case that fits two Jepp binders perfectly, and has a shoulder strap. It cost five dollars at an odd-lot store.

Whatever you use as a flight bag, it should fit your organization scheme. Everything you need to take with you should fit in the bag neatly, and should be accessible in flight.

Information Sources

The most cost-effective source of supplemental information is the Airport/Facility Directory. It has everything you need to actually make any flight, and it's so inexpensive that it's easy to keep a current copy at all times. This should be a required part of your library, since it is the one and only source of "official" information.

Unfortunately, it's missing a lot of information that is particularly

useful, such as FBO listings and locations. For this, there are a
variety of good third-party products. All of these contain extensive
frequency listings, but their drawback is that none are "official"
sources of information, unlike the A/FD. It's tempting fate to use any
of them for critical information on an IFR flight, but they are reliable
enough for VFR use.

• **AOPA's Aviation USA** is a large, thick soft-bound book that
comes along with an AOPA membership. It's also available separately from mail-order houses. It's best for listings of airport businesses, and the one volume contains the entire country, making it a
good buy. It also has a large amount of supplementary information
along the lines of that found in the AIM, and listings of FARs.

• **JeppGuide** was called TannGuide until Jeppesen bought it. It's
an interesting product that is particularly useful for Jeppesen chart
users. Each airport is covered on a page the same size as a Jeppesen
approach plate, with holes punched to fit Jeppesen binders. Unlike
the AOPA book, every airport has a diagram, complete with nota-
tions showing the location of FBOs. The advantage of the *JeppGuide*
is that you can carry along only the airport information that you
actually need. With the appropriate airport's sheet on your
clipboard, you can eliminate some of the preflight work outlined
above. The JeppGuide also has plain-English descriptions of the
airport area and any special requirements for approach and departure.

• **Flight Guide** looks like a miniature Jeppesen binder and packs a
remarkable amount of information into a tiny space. The disadvantage of this is that it's almost too tightly packed. There can be as
many as a dozen airports crammed onto a single 5" x 4-1/4" page, with
a diagram for each. Much of the information is abbreviated, as well.
Still, it's a high-quality product that many pilots swear by. In
addition to airport information, there are Class B/TCA and Class C/
ARSA charts, and listings of navaids. There's also a collection of
other useful information found on the state divider tabs in the book.

• **Air Chart Systems** is a unique product that we'll cover later in our
discussion of charts.

At a minimum, we'd keep on hand a current A/FD, plus one other
guide.

Computers and Avionics

These devices fall into the "optional" category. You can get along just fine without any of them, but depending on your personal style they can be of some use. Those who prefer to keep things as simple as possible will want to avoid these devices altogether: it's easy to let the gadget dominate you in the cockpit, forcing you to serve it.

Many of these devices by their very nature become the centerpiece of a cockpit organization scheme, with the pilot designing the rest of his method around them. Well thought-out systems like this work well, but usually only for the pilot who thought the system up in the first place.

The simplest computer is the tried-and-true E-6B "prayer wheel" that's been in use for decades. It has a number of advantages over newer, fancier electronic gadgets, not the least of which is the lack of need for batteries. It's also much less expensive than an electronic substitute.

That's not to say an electronic flight aid of some sort isn't a good idea. Many pilots today grew up using calculators instead of slide rules, and are more comfortable with the idea of an electronic device than a manual one.

These gadgets fall into four categories: E-6B substitutes, stand-alone flight computers, full-blown laptop PCs running special software, and actual avionics.

• **E-6B substitutes** usually do a bit more than a standard E-6B, but are essentially nothing more than calculators with dedicated aviation functions. Many of the functions are not particularly intuitive, but are not difficult to perform with a little practice. We've flown for years without using anything like these, and haven't missed the functionality a bit—but for those who like to know their wind correction angle, they're useful.

• **Stand-alone flight computers** range in usefulness from utterly awful to superb. We define these gadgets as any electronic device that cannot navigate on its own or be interfaced with an airplane's avionics. The only truly useful ones we've seen are the Flightmaster (the centerpiece of our own system) and Evolution's Pocket Navigator—though this is an expanding market. What makes them useful is an internal database of airports, navaids, intersections and airways; the other units require the user to manually enter latitude and longitude coordinates of every waypoint he or she will be flying over. This is far more trouble than it's worth.

With both the Flightmaster and the Pocket Navigator, the user does a one-time initial setup in which aircraft performance data is programmed into the computer. Thereafter, all that's necessary is to tell the computer your route. It then produces a kind of electronic flight log, which can be printed if desired. The advantage of this over a paper log is that it can be changed at will, very easily—this makes unexpected routing changes a trivial matter.

En route, the pilot punches a button as each waypoint is passed. This automatically updates the log, giving new estimates of time enroute and fuel consumption.

Lastly, the log already contains handy information such as navaid and tower frequencies.

• **Laptop PCs and aviation software** are a relatively recent development, at least in their cockpit-usable form. There have been a number of flight-planning packages available for some time, such as FliteSoft and FliteStar. These also include databases, and produce flight logs, only on paper. More recent developments include graphic flight planning, in which the user indicates his or her route on a chart and the computer figures out the flight log. The disadvantage of these logs is the same as a manually-generated one: if your route changes, they're useless.

By running the software on a laptop or notebook computer and carrying it in the cockpit, you've got the rough equivalent of the stand-alone flight computers described above, albeit with considerably more bulk and expense.

However, the major flight planning programs have an intriguing additional capability: moving map displays. These can be plugged into many lorans or GPS units and will display the airplane's position along with its surroundings on a graphic display that is continuously updated. The average laptop or notebook is a bit large for most cockpits, but some of the software will run on diminutive "palmtop" computers. Some of the packages are even available with built-in miniature GPS receivers. Peacock Systems, makers of the LapMap, has produced products along these lines.

• **Avionics** means lorans and GPS receivers. Today, these are remarkably inexpensive (on a par with comm radios) and amazingly capable. Only a few years ago the pilot was forced to manually enter coordinates for places he or she wanted to fly to. Today, even the lower-end units have full-blown databases, and getting from here to there is as simple as dialing in the identifier of your destination and taking off. For the well-to-do there are even panel-mounted moving map displays that can hook into the loran.

The more sophisticated units have extensive internal information databases including airport frequencies, elevations, and so forth. Some can be removed from the airplane, which not only makes theft less likely, it also allows the pilot to program a flight plan in at home, further reducing the cockpit workload.

With a good loran or GPS unit, any of the other gadgets becomes superfluous. The ability to program in a route lets the pilot dispense with paper flight logs, unless a backup is desired. This assumes that the pilot takes the time to actually do the programming: many pilots we know have lorans with very capable routing functions, yet they usually just plug in the destination and leave it at that.

All of this is moot, of course, if you don't own your own airplane. In that case, you can go with one of the new handheld units, like the Trimble Flightmate or the Garmin GPS-55 AVD.

For those who are willing to put in the time and effort to use these devices properly, any can make navigating much easier and more efficient. Later in this chapter we'll give an example of how an electronic gadget can be used as an integral part of a cockpit organization setup.

Communications

Your workload will go down tremendously if you don't have to strain to hear what's being said on the radio. By all means use a headset. There are several choices available for less than $100 that are perfectly adequate, and any of them beats the cabin speaker hands-down. Comfort is paramount here, so spend some time "tweaking" your headset to fit you. Pilots have griped for years about how the tight clamping pressure of headsets causes headaches. It does, too, unless you spend five minutes carefully bending the headband to fit your head.

By the same token, a portable intercom is worthwhile if you plan on talking to your passengers during a flight.

A handheld radio is a good idea, too. Not only can it provide emergency communications, it can be used to reduce the workload. Before engine start, it can be used to pick up the ATIS without running down the aircraft battery. The same for IFR clearances. In flight, your passenger can use it to pick up an ATIS while you're talking to approach control, to to check weather with Flight Service while you fly.

Miscellaneous Hints

Here are a few tidbits regarding accessories that we've found useful over the years:

One of our favorite inventions is the Post-It. These handy little scraps of paper can be used for anything you want to keep in your line of sight: V-speeds, minimum altitudes, traffic pattern altitudes, winds, runway in use, etc., etc. Be sure to clean up after yourself.

There are many special penlights for pilots, with red lenses and so on. Most are expensive and not worth it. A bright light that uses AA batteries will do fine (AAA batteries are harder to find, and a C or D cell light is too bulky, not to mention that spare batteries are a pain to carry). If you're concerned about night vision, it can be shielded with your hand. We used to use a red-lensed light, but abandoned it. It was useless during night preflights, and we've found the night-vision problem to be nonexistent: white light in the cockpit has never proved to be a problem. For some reason many flashlights are black—this never made any sense to us, since if you drop it you'll want to find it in a dark place. Get one that's a light color.

Every battery-powered device you carry should have at least one spare set of batteries. These should be changed every month, or better still use nicads and recharge them every flight or two. There is no more excuse for having your handheld radio go dead than there is for running out of gas.

We like to use highlighters, but the caps keep coming off in the flight bag. Not long ago we found a dry highlighter in an art supply store; it's a china marking pencil (or "grease" pencil) in fluorescent orange. It does the trick well, never dries out, and costs less to boot.

Don't bother to bring your plotter along. It's useful when laying out a pilotage cross-country on a sectional, but when you start navigating on airways or loran-direct it becomes extraneous. This is particularly true for IFR pilots. We've been flying with Jepp charts for years, and have never had to use the plotter that came with them, not even once.

Paperwork

Flight logs are a little like mouse traps—everyone seems to have a better idea than the next guy. Even the FAA has gotten into the act with its standard flight plan form, which has a flight log on the back.

By "flight log" we mean whatever piece or pieces of paper you use to record information needed during the flight. At one extreme it can be nothing more than a notepad; at the other it might be a commercial one made of plastic, complete with a built-in E-6B, intended to be erased and used over and over again.

In practical terms, we've found that the best logs are the ones you design yourself, because they reflect what's important to you. Later in this chapter we'll look at three specific examples to give you an idea of the shape a log might take.

Regardless of whether you buy, make, or do without a special log, you should keep in mind what it's for—to keep important information in front of you, and to give you a place to record new items. This information might include weather, your flight plan, routing changes, fuel tank times, ATIS, and whatever else you might find important.

Checklists

Every airplane has a checklist, but many are woefully inadequate. Such entries as "Empennage - Check." on a preflight list aren't very enlightening (what are you looking for?).

While the manufacturer's checklist should always be used, we've found it useful to develop our own supplemental lists that are more complete and in many cases more to the point.

There are two ways to go about developing your own checklist. First, you could copy the manufacturer's checklist and insert additional items or expand on those already there. Second, a separate checklist could be worked up to be performed separately from the manufacturer's list.

When developing a checklist, keep it brief but complete enough to be meaningful. You're not writing a whole flight manual, but if you reduce entries down so far that you can't remember what they mean, the list is useless.

A relatively recent introduction is the electronic checklist. Gadgeteers might find them handy, but we prefer the simpler paper variety. To us, the electronic checklist is in the same league as electric carving knives and motorized can openers.

Mnemonics

These handy memory aids can help tremendously in keeping a pilot ahead of the airplane. They're a kind of mini-checklist for oft-repeated items or times when a paper checklist would be too much work. (Remember the rule about secondary tasks during busy portions of the flight? This is where mnemonics come in handy.)

There are many of these, but we'll just outline a few of the better ones here.

• **GUMP(F):** One of our favorites, it stands for Gas (full tank, boost pump on), Undercarriage (gear down), Mixture (rich), Prop (forward, or high rpm), Flaps (as needed). This one is so simple it can be performed any time in a matter of a few seconds. We do it at least three or four times on every approach or trip around the pattern. It keeps the important things in mind, like lowering the landing gear— we even do that in fixed-gear airplanes, just to stay in practice. (We know of one pilot who used that step to look out the window of his Cessna to make sure the gear looked okay—one day, he found that the left main wheel had fallen off. Had he not done a gear check, he would never have known and might well have become a statistic.)

• **Five As:** For departure and arrival, this stands for: Atis, Altimeter setting, Avionics set, Airspeed to use, Approach plate available.

• **Six Ts:** For IFR pilots, it means Turn (to the new heading), Time (start the approach timer), Twist (the OBS to the inbound heading), Throttle (down to approach power setting), Talk (to the next controller), Track (the new heading). It's for holding and full instrument approaches, but it can be used enroute as well, since controllers often hand off to the next facility at the same time the pilot passes over a notable waypoint, like a navaid. In enroute use, of course, Time would be to note passage of the fix, and Throttle would not be used unless an altitude change was called for. VFR pilots could use this when passing fixes on their route as well, to keep them on top of where they are and what they should be doing.

• **ISHAFTM:** Another one for IFR pilots, it means Identify (the correct approach chart), Speed (to fly the approach), Heading (inbound), Altitudes (minimums for the approach, last cleared altitude, altitude for the missed approach), Frequencies (on all comm and nav radios), Time (start the approach timer), Missed (review missed approach procedure).

• **MARTHA:** Another IFR approach aid, it goes like this: Missed Approach, Radio frequency, Time, Heading, Altitude.

• **CRAFT:** This is a clearance mnemonic that works for IFR pilots or VFR pilots who get clearances at large airports. It follows the format in which clearances are delivered: Cleared to via Route at Altitude, contact departure on Frequency and squawk a Transponder code.

Some pilots like to put the mnemonic on their flight logs so they won't forget to run through it occasionally.

Another good technique to use along the lines of memory aids is to say everything out loud. Verbalizing an action will serve to call more attention to it, and will help prevent you from getting distracted. Don't think this is silly: the best pilots in the world do it. During the approach to the Apollo 11 moon landing, Neil Armstrong and Buzz Aldrin kept up a running commentary all the way to touchdown. It kept the important things—speeds and rates of descent—in focus at a critical and highly stressed time.

Charts and Plates

These are really the centerpiece of your cockpit information. You should at a minimum, carry current version of all the charts appropriate to the area and your qualifications.

There's really no excuse for not having a current chart of the area you're flying into. Even at six dollars a copy for sectionals, they're cheap compared to the cost of operating an airplane in the first place. Also, you're guaranteed of having the latest information on ATC frequencies and airspace boundaries. There's no need to compulsively go out and buy the new edition of every chart you're ever likely to need, or to subscribe to them. If your local FBO hasn't got a needed chart in stock, you can call a mail-order house. These businesses carry everything you'll ever need, and if necessary can get it to you overnight with a single phone call.

If you're flying VFR into an area of Class B airspace, the appropriate terminal chart should also be secured. It provides an enlarged view of the complex Class B area, along with useful notes not found on sectional charts.

WAC charts may be useful in the wide-open spaces of the West, but in a crowded area they're too dense to make much sense of.

If you're a VFR pilot, you should consider carrying IFR charts of the area in addition to sectionals. These are actually maps designed to show navaids and airways clearly at the expense of geographic detail. If you navigate by navaid instead of pilotage, an IFR enroute chart can come in handy. By the same token, instrument pilots should carry a current sectional along—it has information on it not found on IFR charts, and it can prove very useful.

As noted above, when you're planning your flight, you should go over the chart even if it's a trip you've made many times before. Things change, which is why new charts are issued in the first place.

A book of approach plates is also a good thing for a VFR pilot to have. You may not be instrument-rated, but the plates can provide you with a lot of good information in an easy-to-read form. This includes frequencies, detailed diagrams of larger airports, and for major areas details of special visual approach procedures.

There are three major chart systems: NOS government charts, Jeppesen-Sanderson, and Air Chart Systems. With the exception of special VFR charts for the L.A. basin, Jeppesen does not produce charts for VFR pilots. The other two publishers produce both VFR and IFR charts.

Air Chart Systems' product is actually a repackaged version of NOS charts combined with a special VFR chart format and supplemental information in tables. There are some advantages to these charts, including cost-effectiveness and their format (all charts are in large spiral-bound books that are easy to handle in the cockpit). Many pilots like them, but we've found some disadvantages. For one thing, the reprinted charts are in a single color, which in the case of reprinted NOS sectional and TCA charts causes much of the detail to be muddied. There's a VFR atlas, an IFR atlas, and a terminal directory. Air Chart Systems has a money-back guarantee, so if you're curious you can try them with no risk.

Most VFR pilots will opt to go for NOS charts. If a VFR pilot wants to carry IFR charts, it makes more sense to use NOS than Jeppesen, since NOS charts and approach plates can be bought individually, while Jeppesen products are available only through subscription. These subscriptions are expensive, and not really justifiable for the typical VFR pilot.

The majority of IFR pilots choose between Jeppesen and NOS, and the argument as to which is better is an old one. Essentially it comes down to personal preference.

Three Systems

Let's take a look at three different methods for keeping yourself organized in the cockpit. In each case, the pilot who designed the method is instrument-rated, but each method could be easily adapted and used by a VFR pilot.

One of the three is simply a well-designed flight log sheet sized to fit in a Jeppesen binder. The second uses a handheld flight computer, the Flightmaster, as its centerpiece, and the third is a highly personalized set of supplemental checklists.

• **Jepp-sized approach aid.** This is a clean, simple form specifically designed to assist the pilot fly instrument approaches. The idea

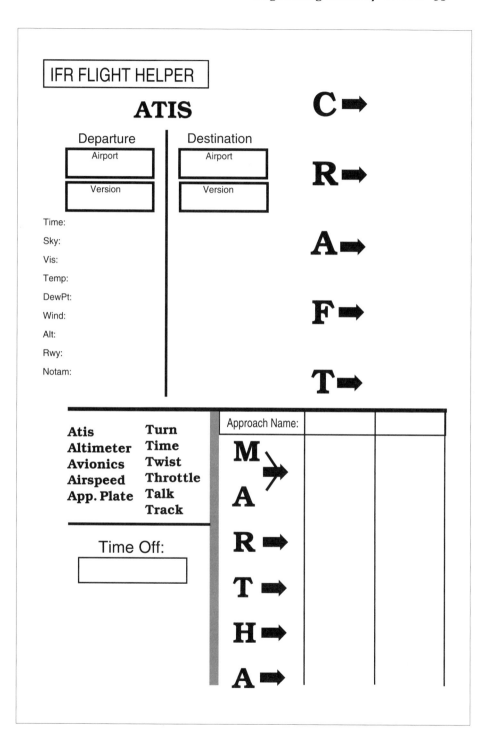

IFR FLIGHT HELPER

ATIS

C→

Departure
Airport

Version

Destination
Airport

Version

R→

Time:

Sky:

Vis:

A→

Temp:

DewPt:

Wind:

F→

Alt:

Rwy:

Notam:

T→

Atis   **Turn**
Altimeter **Time**
Avionics **Twist**
Airspeed **Throttle**
App. Plate **Talk**
    Track

Approach Name:

M↘

A

Time Off:

R→

T→

H→

A→

here is to provide a memory aid during the busiest portion of the flight, thus reducing cockpit workload at this critical stage.

The form presents the five As, six Ts, CRAFT and MARTHA mnemonics designed for IFR pilots (see the section on mnemonics earlier in the chapter). Note that two of the mnemonics are actually spelled out in checklist form. There's also space for the ATIS at both departure and arrival airports.

This pilot also would carry a notepad for other information, routing amendments, and so forth. In this case, the only information he feels he needs to organize are the items on the sheet. Everything else can be written down as he receives it. This keeps thing simple for this pilot.

• **Flightmaster-based system.** This system is relatively complex, and is set up to account for all parts of a flight. It's centered on a small flight computer called the Flightmaster, which is used to do all the calculation work and much of the record keeping for the flight. The pilot actually does relatively little in the cockpit, leaving him free to fly the airplane, deal with ATC and watch for traffic.

The Flightmaster, when it was introduced, was an unusual product. Based on a small British handheld computer called the Psion Organiser, it differed from previous flight computers in that it didn't function merely as a special-purpose calculator. Instead, it was designed as a kind of electronic flight log that would automatically calculate everything from fuel burns to wind correction angles to weight-and-balance and present it to the pilot in a form that was very usable in the cockpit. Not only that, but when used in flight it would automatically update everything as the pilot crossed each waypoint on the flight plan.

The Flightmaster also included a full database, which is what made it really useful. Previous handheld nav computers were limited in that the pilot was forced to manually enter coordinates for every desired waypoint on a route—tedious at best. With the Flightmaster, the pilot could simply key in the identifier for the starting airport, any victor airways and/or navaids along the way, and the destination and it would generate a flight plan and trip log on the spot.

This device replaces the paper flight log student pilots are so familiar with. It provides all the information a paper log does, plus frequencies of navaids along the way. As a result, part of the paper form for this system that had originally been used as a log is now available for use as a note-taking area.

Out:	In:	clock time	clock time	clock time	clock time	clock time
Start time:	Takeoff time:	left / right	left / right	left / right	left / right	left / right
on L / R	on L / R	elapsed time	elapsed time	elapsed time	elapsed time	elapsed time

Amendment #1			Altitude		Frequencies	
Who					Who	Freq.
Time						
Location						
Amendment #2						
Who						
Time						
Location						
Amendment #3						
Who						
Time						
Location						

Approach Notes for:		RV:				
Freqs	VOR:					
ATIS:	ILS:					
Tower:	LOM:	**DH**				
Ground:	Missed:	**MDA**				
		GS and Time				

N-

is cleared to

void time

via   .

As filed?

Fly	Climb / Maintain	Expect / in	ATIS - Departure	ATIS - Arrival
Rwy Hdg?				
Frequency		Squawk		

Synopsis:

	Departure ID:					Destination ID:
Reported						
Forecast						
Winds -30 / 60 / 90 / 120						
FZLV, Turbulence, Fcst Tops						
NOTAMS						
Pireps						

1. Type	2. Tail Number	3. Equipment	4. Airspeed	5. **Departing**	6. **Time**	7. **Cruising**

GMT=+5,+4DS

8. **Route of Flight**

			Weight	Moment	Limits:
Fuel 48.43"			—	—	BEW=1837, cg 45.0
P. seat 32.75"		155	5076		Moment 82665
CP. Seat 32.75 - 39.0"		—	—		Gross = 2740
Bk. Seat 70.7"		—	—		@gw, 45.0-50.1
Luggage 95.5"		—	—		@2480, 41.8-50.1
Hat rack 119.0"		—	—		@2250, 41.0-50.1 Below 2250 same

9. **Destination** | 10. **Time Enroute** | 11. Remarks

12. **Fuel Time** | 13. **Alternate** | 14. Pilot | 15. **SOB's** | 16. Color

c.g = _____

The weather log used with the Flightmaster-based organization scheme. At full size it fills an 8-1/2 x 11 inch sheet.

The other components of this system are Jepp charts and plates, sectionals, a letter-sized clipboard, a set of customized checklists, a weather data sheet and the information log on the preceding page.

This is a compartmentalized organization system, with two forms that are filled out for every flight. The pilot in this case likes to have a specific space to record all sorts of information, unlike the one who designed the approach aid described earlier.

The weather sheet has a space across the top for the synopsis, which is the first thing a briefer will give the pilot. Underneath this is a table with a row for each type of information given in a briefing, such as current and forecast weather, winds aloft, pireps, and so on. There are several columns: the leftmost one is used for weather at the departure airport, the rightmost for the destination. The columns in between are used for weather at sites en route. At the bottom is a reproduction of a standard FAA flight plan form and a space for weight and balance calculations.

The flight data sheet occupies half of a sheet of letter-size paper. This is shown on page 97. The other half was originally used for a standard flight log, but with the Flightmaster that space is used for notes. Most of this sheet remains blank until the trip is started. Frequencies that will be needed along the way for VFR are written in the notes area (when IFR, frequencies are supplied to the pilot in flight).

At the top of the sheet is a row of boxes to record tank-switching times. Below this are several boxes to record ATC routing changes for IFR, or ATC instructions for VFR. Each box has notes to remind the pilot of needed information. To the right of these is a set of boxes to record radio frequencies, which facility they're for, and the current cleared altitude. This serves as a communcations log: in case the controller on the next frequency down the line doesn't answer, or if a mistake was made, the previous frequency is recorded and the pilot can go back to it.

Underneath the routing changes boxes is an area that is used to record pertinent information for the approach to be used when flying IFR, including frequencies and altitudes.

The bottom third of the sheet is for clearances. It is used for every IFR flight, and for VFR flights out of airports in Class B or C areas where departure clearances are issued for all flights.

The checklists (page 100) are copies of the factory checklists with additions. They were resized to fit in Jeppesen chart protector sleeves. The appropriate approach plates are also placed in sleeves, and the whole assembly is held together with looseleaf rings. This makes a kind of book that is convenient to use and has everything that's needed during the flight.

In use, the charts and plates for the trip are pulled out of the Jepp binders, with the plates going into the checklist "book." A route is selected, and plugged into the Flightmaster, a process that takes a couple of minutes. The flight plan form is also filled out. A weather briefing is obtained, and the weather chart filled out. The winds aloft are plugged into the Flightmaster during the briefing to get an accurate time enroute estimate, and the flight plan is filed.

Once everything is assembled and packed, it's off to the airport. In the cockpit, with preflight done, clipboard out and ready and the flight bag stowed, the ATIS is copied and a clearance is obtained. There's usually a change from the filed plan, so it's written down, plugged into the Flightmaster and marked on the chart.

Before the engine has even been started, this pilot is completely set up for the flight. The route is set, has been marked on the chart,

RUNUP

Parking Brake — SET
Selector -- FULL TANK
Mixture -- RICH
Throttle — 1900-2000
Mags—175 DRP, 50 DIFF
Engine inst ------GREEN
Vacuum light --- OFF
Throttle-- 1500 RPM
Prop— CYCLE
Throttle — 1000 RPM
Controls—FREE
Radios, instruments — SET
     NAV1: 1ST FIX
     NAV2: 1ST INT
Strobes — ON
Annunciators, AP — TEST
Trim — TAKEOFF
Flaps — 15 DEG
Door — LOCK
Harnesses — SECURE

TAKEOFF

Fuel Pump — ON
Transponder - ON-ALT
Timer, time off
Power—FULL-2700 RPM

ROTATE AT 71 MPH
CLIMB AT 82 MPH
ACCEL 105-115 MPH
(SHT. FIELD 76 CLIMB)

Gear — UP (125 mph max)
Flaps —UP
Fuel Pump-OFF, PRESS

CLIMB
Power — 26 IN. 2600 RPM
Mixture-LEAN as needed
Cowl Flaps — OPEN
Airspeed — 105-115 MPH
Ram Air — ON in clear air
WATCH MAN PRESS.

DON'T FORGET !
COWL FLAPS
FUEL PUMP
MP 24- 24

Vy - 100 MPH
Vx - 76 MPH

Va - 138 MPH
Vno - 200 MPH

For DESCENT:
VGear
150 MPH - 23.6" - 2200 RPM
VFlap
132 MPH - 20" - 2200 RPM

For APPROACH:

As needed -- appch
105 MPH
Final 81 MPH

One of the custom checklists used with the Flightmaster-based organization scheme.

and the flight log is done. The appropriate charts are in hand and ready.

For runup and departure, the customized checklist is used, then put out of the way once established enroute.

Enroute, weather updates are put on the weather sheet in unused columns. The only thing the pilot needs to do is tell the computer as each fix is passed, and record frequencies, altitude assignments and cleared altitudes.

Before an approach, the pertinent information is recorded on the sheet, the charts are put away, and the approach plate put on the clipboard. The Flightmaster is also stowed. Everything is thus out of the way for the final, busy portion of the flight.

• **Specialized Checklists:** This pilot spends most of his time flying IFR, and has devised a set of highly specialized checklists to be used in conjunction with the aircraft checklists to stay ahead of the airplane. The information sheet is very simple, with a space for the flight plan, ATIS, clearance and amendments. This pilot also carries a notepad for written information.

These lists are cryptic, which lets the pilot fit more on a sheet of paper. However, that makes them less understandable for others; this is one reason that these systems work best if you design them yourself.

The lists are shown side-by-side on the next page. Normally they'd be printed on separate sheets of paper. There are two lists, one for departure and immediately after takeoff, the other for approach. This pilot prefers to use only a notepad enroute.

The departure checklist has reminders at the top of what is to be done during the planning stage. Below that is a cockpit checklist to be completed prior to departure. There are also reminders of tasks to be performed as the airplane passes through 1,000 feet on departure.

The arrival list uses the ISHAFTM mnemonic, spelled out. There's also space to write down the ATIS, missed approach information, and a variety of other bits of useful information.

There are a few advantages to splitting up the sheets this way. Since each sheet deals with a different aspect of flight, so it can be stowed when the pilot is done with it. By the time the approach comes along, all that's left is the plate and approach checklist, again keeping things as simple as possible in the cockpit.

The pilot who devised these lists also makes extensive use of techniques like saying actions out loud, using Post-It notes and

IFR Check List

Flight Plan:—Weight and Balance—Weather
•Fronts (Check opposite side)
•Convection
•Ice
—Route—Alternates (P-600/2;N-800/2;
None-1hr.,2000ft.,3mi.)
—Fuel Requirement/Burn—Runway Lengths
—NOTAMS—File—ATIS—Clearance

Remember: Taxi Checks, GTPFBCM, 6Ts

IFR Cockpit Check
—Transponder •Squawk •Test •Standby
—VOR #1 & #2•Check with local facility
•Prepare #1 and #2 (FAF?)
for return or 1st fix, 1st intersection
—ADF
•Tune and ID to compass locator
for return to departure point
—Comm
•Tune and transmit #1 and #2
•Check Handheld
—Markers
•Check lamps•Audio on•Switch Hi
—DME
•Check
—Taxi Checks
•DG•TC•AI
—Run Up
•Review SID, Return Approach
•Review IFR Departure Procedure
—BLTT
•Boost•Light•Transponder to Alt •Time Off
—Lost Comm
•ACE Assigned, Charted, Expected (Altitude)
•ADEF Assigned, Dir, Expected, Filed (Route)
•7700/1min.; 7600/15min.;Repeat
—VOR Check
•DEPS Date, Error, Place, Signature
—Reaching Hold
•FAT Fix, Altitude, Time

Approach Check List - 6Ts

ATIS

ISHAFTM

I - Identify correct approach chart
S - Select a speed to fly the approach
H - Select an inbound heading
A - Review altitudes, memorize minimums, MSA,
Post It
F - Frequencies on all nav/comm aids, markers
ADF, DME
T - Zero the clock and note time to missed
M- Missed approach procedure

6Ts

No. 2 VOR Shift to Inbound Freq/Heading if no
radar available
Shift to IAF After Missed if radar
available

Approach Lighting Review

—How Low _____Ft.
—How Long_____Time
@ GS_____Kts. VSI _____Fpm
Missed Approach
Climb_____ To _____Ft.
Turn_____ To _____Ft.
Heading_____Deg.

The two primary checklists used in the supplemental checklist method. The left one is used prior to takeoff, the right for approach. Enroute a simple notepad is used. The cryptic "mnemonic" **GTPFBCM** *is a reminder for a post-takeoff check: Gear, Throttle, Prop, Flaps, Boost pump, Cowl flaps, Mixture.*

using cockpit instruments as memory aids. All of this keeps him well ahead of the situation at all times.

Summary

The techniques outlined above give you a glimpse at several very different approaches to keeping a cockpit uncluttered and a pilot in charge of the flight. We're not suggesting you adopt any of them wholesale: rather, they're intended to give you some ideas on how to set up your own scheme.

We do recommend that you develop a method for organizing yourself, and stick with it. You may not need it for every flight, but by staying in practice it will serve you well when things get hectic.

Above all, you should be comfortable with whatever you choose to do to stay organized. Don't try to adapt your habits to someone else's system; adapt the system to fit you.

6 Small Airports

In a way, operating from small, uncontrolled airfields is more dangerous than flying into Atlanta Hartsfield. Not only can there be a very dense mix of traffic, there's the added complication of a lack of air traffic control. It's entirely possible to have a pattern chock-full of aircraft, some of which have no radios, mixing it up with pilots practicing instrument approaches to the airport—all without the benefit of a controller to keep everyone from running into one another.

The small, untowered airport lies at one end of the GA airport spectrum. The other is occupied by primary airports inside Class C/ARSAs, and reliever airports located near major metropolitan areas—places like White Plains, New York's Westchester County Airport, New Jersey's Teterboro, or California's John Wayne in Orange County. Operating at these fields is much like flying into large airfields, but without large jets to contend with—they can be very busy, indeed.

Uncontrolled Fields

Frankly, we'd rather fly into the beehive that's Washington, D.C. than try pattern work at a popular non-towered airport on a fine summer day. The problem is lack of organization, and that can be dangerous. With nobody to call the shots, there's no telling what might happen.

Naturally, if all of the pilots would follow the rules precisely all of the time, things would go smoothly...and they often do. The problem is that many pilots play fast and loose with the rules, thereby creating hazards for all. "See and be seen" takes on its full meaning in such an environment.

Non-towered field operations have built-in hazards even for those who do everything right. Once we were flying into Frederick, Maryland, the uncontrolled field that's home to AOPA. The weather was about 900 overcast, and clear below, with runway 5 in use. Runway 5 does not have an instrument approach, but the reciprocal, runway 23, has an ILS. We were cleared to shoot the approach, and were intending to circle around to runway 5 after breaking out. We came down the ILS, announcing our position on unicom, and broke out right where we were supposed to be...and found ourselves eyeball-to-eyeball with a no-radio Cessna 140 doing pattern work. We never got closer to one another than a quarter-mile, but it was unnerving nonetheless. The disturbing thing was, we both had every right to be there.

While we encourage pilots to take advantage of the help that ATC can provide when flying in crowded airspace, the area immediately surrounding an uncontrolled field is one place where ATC is incapable of rendering any assistance whatever. It's entirely up to the pilot to keep himself safe in this environment.

There are two keys to this: first, to keep a *very* sharp lookout whenever near a non-towered airport: just monitoring unicom is not enough to keep track of the traffic. The second is to communicate properly and effectively.

Keep a Lookout

When at a towered field, it's not really necessary to keep aware of everyone in the pattern. The tower controller does that for you. At an uncontrolled field, however, it's up to you to determine when it's time to take off, whether or not it's safe to land, and if your spacing is adequate in the pattern. You need to be aware of everyone operating at that airport at the same time you are.

When departing from an uncontrolled field, you can begin this process by paying attention to the pattern and runway operations during your preflight and taxi. Often, by the time you reach the runup area you'll have a good idea of where everyone is in the pattern. Still, there may be aircraft with no radios arriving, or you may have missed someone. One of the last things you do before taking the active should be to do a 360 in place on the taxiway, looking for traffic in the pattern.

This is another good place to get your passengers involved. The extra sets of eyes are invaluable in looking for traffic. Be sure to talk to them about the traffic, rather than assuming you're both looking at the same airplane.

In the pattern, keep track of where everyone else is. Again, it's far more important to do this at a non-towered field than it is at a controlled airport.

Also, when in the pattern it pays to watch the approaches to pattern entry closely. That includes not only the correct one (45 degree entry to a midfield downwind) but also the crosswind, base leg entry, and straight-in approaches. Many pilots use these, even though it's unsafe to do so. Further, be aware that some pilots may use the other pattern (right hand vs. left hand), especially at airports where nonstandard patterns are preferred.

When departing, it's a good idea to use a standard pattern exit (45 degree turn away from the runway heading until clear) and to climb above the pattern as quickly as possible.

When arriving at a non-towered airfield, you should always use the standard pattern entry if at all possible. Be aware that departing pilots tend to just head out in whatever direction they please, and often leave the frequency immediately after departure...so it's entirely possible to have aircraft popping out of the pattern at random places, heading off in random directions, without the benefit of radio contact.

If there are aircraft in the pattern, it's a good idea to spot one, identify which it is, and keep it in sight as you take up position behind it. As long as you can see the other guy, you can be sure he won't have the chance to run into you.

Communications at Uncontrolled Fields

The way many pilots conduct themselves on unicom frequencies is, to put it mildly, terrible.

It's all too common to hear things like "Cessna's on downwind," or "Hiya, Joe, what's going on with you?" over unicom at untowered fields. Which Cessna? Where on downwind? Which airport, for Pete's sake? And why can't that pilot just talk to Joe after he lands and get the heck off the radio?

Your safety at crowded uncontrolled fields can be greatly enhanced by using the radio effectively. That means short, concise transmissions only when necessary, and those transmissions should include only needed information.

Your general style of radio usage should be somewhat different on unicom than it is when dealing with ATC, simply because you're conveying different information. Most of your transmissions will be self-announcements, along the lines of 'here I am, and this is what I'm doing.'

Any self-announce transmission should include, at least, the name of the airport, your call sign (even if abbreviated), the runway number, and *exactly* where you are. It's a good idea to end the transmission with the airport name as well.

The reason for this is simple. There are only a few unicom frequencies, and they're so congested with radio traffic at different airports that a transmission heard in full without being "stepped on" is the exception rather than the rule in many parts of the country. By including the airport name at beginning and end there's less likelihood of your location getting lost in the babble.

A typical transmission should go like this:

"Oxford traffic, Cessna 1234 midfield left downwind runway 18, Oxford."

That gives anyone on the frequency everything they need to know: which Cessna you are, the fact that you're not the guy about to turn base, which side of the field you're on, which runway you're using, and which airport you're at.

Self-announcements should be made when approaching the field, on downwind, turning base, turning final, on short final, on crosswind, and when clear of the airport area.

When approaching and departing the airport, try to avoid making transmissions at high altitudes. Since VHF radio is line-of-sight, tuning in unicom at 3,000 feet can easily result in a hopeless mess of stepped-on radio transmissions and static. By dropping down to pattern altitude a few miles out, then calling, you'll have an easier time of contacting unicom at the airport you're operating into.

One important way in which transmissions at uncontrolled fields differ from those to ATC is in the ability to provide helpful information to those monitoring the frequency. Some examples:

• If you're entering the pattern behind a Cessna pilot who just called midfield downwind, you can self-announce and add that you have him in sight. This helps everyone in the pattern find you, and takes the heat off the pilot you're following since he now knows you're not about to ruin his day by colliding with him.

• When approaching the field, rather than just calling for runway in use and wind advisories, throw in your location, including altitude, for the benefit of those in the pattern. If they know you're still ten miles out they won't have to worry about you blasting into the pattern in the next ten seconds.

• If conducting practice approaches at an uncontrolled field, make self-announcements at regular intervals. Also, don't assume that everyone knows what you mean when you use IFR terminology. Telling a pilot that you're marker inbound on a practice ILS won't do any good if the fellow in the pattern doesn't even know what a marker beacon is, much less where it's located.

• Convey your intentions, so people will know what to expect of you. If you're staying in the pattern, say so. Let the other pilots know which direction you're headed if leaving the pattern—this alerts arriving traffic that you may be headed their way.

• On the ground, many pilots are in the habit of calling unicom for a radio check. It is a good idea to verify your transmitter's operation, but why not combine that with some useful information that might benefit another pilot? Rather than just asking for a radio check, ask for the winds and runway in use.

Special Considerations at Small Airports

Many small airports are home to aviation-related businesses, like soaring or skydiving operations, banner-towing outfits, large flight schools, aerobatic schools, and the like.

If you're headed to an unfamiliar field, this information can be vital. You'll be busy enough trying to watch for traffic in the first place without having to worry about cropdusters flying at treetop height.

Sometimes the sectional chart can provide a clue to operations to watch for in the area. Skydiving outfits are indicated by a parachute symbol, and soaring is indicated by a glider symbol.

There's no indication of other types of activity, however. Often, this can be localized knowledge not published anywhere, but of great importance to pilots flying in the area. For example, at Allaire airport in Belmar, New Jersey there's an active banner-towing business that essentially runs all the time in the summer months. The aircraft operate at low altitude and low speed, and shuttle back and forth between the airport and the beach several miles away. These pilots do this all day long, and don't always announce their location. If you fly into this airport you learn about this, but it's not published anywhere. The only way to get this kind of information is to call the airport ahead of time and ask.

ATC and the Uncontrolled Field

By definition, the services of air traffic control are not available to pilots at uncontrolled airports. Once the pilot climbs into controlled airspace, however, he or she may be able to take advantage of services like flight following on VFR flights, a practice we heartily encourage.

The question is, how does one transition from the uncontrolled environment of the non-towered airport into the system?

For an IFR pilot, that's simple. If a VFR departure from the airport is not possible, an IFR pilot is issued a clearance with a "void time." The pilot has to get airborne and contact ATC before the void time or the clearance is invalid. On the other end, ATC releases the pilot once the approach is commenced.

For the VFR pilot, it's a question of knowing who to call for ATC service in your area. For most of the country, the controlled airspace above a non-towered airport "belongs" to an approach control facility. If the airport has an instrument approach, the frequency will be printed on the approach plate. Airport directories also usually have approach frequencies. Often the frequency listed isn't the one for the pilot's exact location, but the controller will provide the right one—be prepared to copy it on your first call.

When arriving at a non-towered airport, a VFR pilot should cancel flight following fairly far out—say 15-20 miles—and begin a descent to pattern altitude. This does two things: first, it lets the pilot begin monitoring unicom early to get a picture of what's happening at the airport. Second, letting down early reduces the problem of congested unicom frequencies. What is unintelligible at altitude due to congestion may be crystal-clear at 1,500 feet.

Non-reliever Towered Airports

In this category we include controlled fields that are not in close proximity to major airports. Those we call relievers and are covered under a separate heading below.

These are among the easiest airports to fly into in the country. While busier on the average than most uncontrolled fields, having a tower controller to help out with traffic avoidance is a big safety benefit.

Typically, these airports do not have jet airline service, but they often are host to one or more commuter airlines. Usually the volume is so low that the average pilot hardly notices them. The biggest activity at these airports is often training. If you fly out of or near one of these fields, it's a good idea to learn where the practice area is and keep clear of it.

It's also wise for the VFR pilot to take a look at the instrument approaches into these airports. Not only are they often used for practice, many pilots fly IFR even on VFR days and will be coming in from predictable directions.

Relievers

These are the largest and busiest of the "small" general aviation airports. They're used as alternatives to major metropolitan airports, and are favored by regional business travelers and corporate flight departments.

As a result, they can be every bit as busy as a major airport, without the benefit of being surrounded by a Class B/TCA or Class C/ARSA. Usually, though, they lie under the edge of such an area or just beyond its perimeter.

"Reliever" is not an official definition, and there's no real dividing line between a reliever and any other towered airport. It's safe to assume that if it's near a large city, it'll fall into this category, however.

Because of the heavy traffic into and out of these airports, it's unlikely that there will be a lot of training activity going on. While it's possible to do pattern work at these airports, it's a better idea to go and find a quieter field.

A good example of a reliever airport is Westchester County in White Plains, New York. It's one of two major relievers for the New York TCA, which has three primary airports in it to begin with (the other is Teterboro). We'll use it as a case history to illustrate the kinds of problems a pilot is faced with when flying into such a field. Mostly, they come from volume of traffic and the geography of the surrounding area.

Westchester is a two-runway airport that lies under the uppermost "deck" of the New York Class B/TCA. As of this writing, the upper edge of the White Plains ATA butts against the floor of the TCA. When the ICAO airspace designations go into effect it's likely that FAA will either drop the floor of the Class B area or raise the roof of White Plains' Class D area to keep a slot from opening up over the airport.

To the south is New York City and its Class B/TCA, which effectively forms a 'wall.' It's unlikely ATC will let you into the TCA

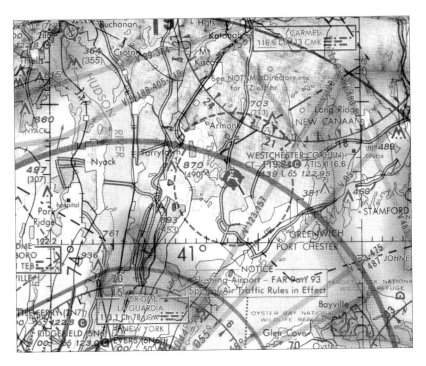

White Plains, New York (HPN) is a well-known reliever airport for the New York TCA. None of its quirks or hazards are shown on the chart, though a close look can help a pilot unfamiliar with it deduce most of them.

south of White Plains, since just beyond the edge is LaGuardia. To the west is the Hudson River, which happens to lie just about five miles from the field. Over the Hudson there's a VFR corridor that goes right through the heart of the Class B/TCA, which means a lot of VFR traffic operating low over the river. To the southeast is Long Island Sound. There's also a "hole" under the TCA deck south of the airport's ATA and north of the TCA "stem," making it possible for VFR pilots to get from the river over to the sound.

To the north is rolling terrain and Carmel VOR, over which much of the inbound IFR traffic is routed. To the east is the city of Stamford, where VFR traffic inbound from the east congregates.

The IFR runway is 16/34, so IFR traffic is funneled in towards White Plains from the northwest or southeast, depending on the wind.

Essentially, the airport's ATA is completely surrounded by heavy traffic, airspace restrictions, or both. This can be expected at any reliever airport.

The airport, itself, can handle aircraft as large as DC-9s and 737s. It's also home to the jet fleets of companies like Xerox and Pepsico. As a result, there's a lot of turbine traffic and airline operations, but no heavy jets. Departing jets aren't much of a problem for GA pilots in the area, since they climb up into the TCA very soon after takeoff. Arrivals are another matter, though. The controllers at White Plains are presented with the daunting task of mixing Skyhawks, Lear jets and Beech 1900s on a regular basis, and will often advise that there's a Lear right behind you coming down the ILS—something guaranteed to make any pilot unused to such dense traffic more than a little nervous.

Being one of the closest GA airports to New York City, there's also a lot of GA traffic. There are even a couple of flight schools on the airport.

White Plains is somewhat unusual in that it's necessary to contact New York Approach for permission to enter the airport's ATA. This fact is not published on the chart.

The approach control frequency is what you would expect, very crowded and busy. It's best to call from a convenient and well-marked location just clear of the ATA, like a nearby VOR—go there and hold while you raise the controller.

A look at the standard instrument departure for the airport gives a clue as to where the traffic will be. IFR departures to the north and east are sent over Carmel VOR to the north. Those to the west and south are sent over Sparta VOR to the west, which lies just north-west of the edge of the TCA. Sometimes departures are sent south over JFK. It's unlikely that many people will be arriving form the south, however—most IFR arrivals are brought in over Carmel to the north or Sparta to the west.

Also, pilots can count on traffic skirting the edge of the TCA and ducking under it. to avoid having to deal with ATC.

Westchester is near a coastline, and a good rule of thumb is that there will always be traffic flying up and down it, everything from traffic reporters to VFR pilots going up the coast to airports in Connecticut.

One last significant wrinkle to this airport's surroundings is that pilots inbound from the east are usually asked to hold over the city of Stamford, Connecticut. From there the airport is semi-hidden behind a low ridge, and in the afternoon with the sun behind it it's virtually impossible to see.

You can see how this is a tough airport to visit. Very crowded, very hectic, and with dangerous surroundings in terms of traffic and airspace traps that can get you busted.

How best to operate there? First of all, don't count on being let into the TCA, either arriving or departing. Still, it's worth a try, particularly departing to the south or west. A call to clearance delivery might get you a TCA clearance, which will get you into the system and relative safety.

When departing the area, by all means get away from there as fast as possible. Be very careful of aircraft skirting the edge of the TCA and ducking beneath it. Keep your altitude low to keep from climbing into the TCA by mistake. If you're IFR rated, by all means file. Getting into the system is much safer that trying to duck around its edges. If you're VFR, however, you may find that flight following is tough to come by this close to the TCA—another reason to try for a TCA clearance on departure.

When flying into a reliever like this there's no substitute for advance planning, especially the first time. Read up on the airport and study the chart, and devise a specific plan for arrival. The important thing is to avoid arriving "dumb," without a clear idea of what to expect.

Arriving from the north, northwest or east it's best to just duck under the edge of the TCA, since the airport is so close to its northern perimeter. From the south or southwest, you must get around or through the TCA somehow. To be safe, don't rely on ATC allowing you through it. Instead, you can plan to use the Hudson river VFR corridor or just count on skirting the TCA to the west and north. Again, these kinds of strategies work at all relievers, but the details will be different.

Communications at airports like this are much like they are in any busy ATC environment. If you treat the controllers at busy relievers like those at a large TRACON, you should get along just fine.

Class C Primaries

A large number of major airports in the United States don't warrant being surrounded by a Class B/TCA. These are airports at which the major activity is airline operations involving aircraft of all sizes, including large jets.

Individual ARSA primaries vary widely from one another. Some are relatively sleepy and easy to operate at (Portland, Maine is a lot less hectic than White Plains, New York, for example, even though

Portland is an ARSA primary), while others are so busy they might as well be Class B/TCA airports.

While the rules inside a Class C/ARSA are less restrictive than those inside a Class B/TCA (see Chapter 3), it's a good idea to treat them both equally. Simply be aware that you're dealing with a controlling agency, in this case the TRACON for the primary airport, and that you should do what they tell you to do.

Often, you'll find that flying into an ARSA primary is easier than going to a reliever. This is because relievers are close to or underneath Class B/TCAs, and have a lot of traffic surrounding them, but without the benefit of positive control that the Class B/TCA provides. Flying into an ARSA primary, on the other hand, is much like flying into a Class B/TCA primary, only far less intense. You have ATC giving you traffic advisories if VFR, making your life much easier as a pilot.

When operating into a Class C/ARSA, you're supposed to contact ATC 20 miles out. It's a good idea to pick up the ATIS from much further than that, then monitor the frequency for a while to get an idea of the traffic volume before calling in. A side benefit of using flight following crops up here: if you're already being handled by ATC, the controller will simply hand you off just as if you were on an IFR flight, with no interruption of service.

When leaving a Class C/ARSA primary, you'll need to contact clearance delivery for a transponder code and departure control frequency. Other than having to talk to departure on the way out, it's essentially the same as flying out of any towered airport.

7 | Large Airports

I t is interesting that many of the large airports like Chicago's O'Hare and Miami International that are feared the most by general aviation light plane pilots are actually among the easiest to operate in and out of in the country. This is for several reasons.

The first is that these airports are set up to handle large volumes of air traffic. It is *normal* for the tower and ground controller at O'Hare to be dealing with twenty or thirty pieces of traffic at the same time. If the same thing were to happen at Tallahassee Municipal you would see some problems develop because they aren't trained or equipped to handle such a load. At large airports, being busy and overloaded is the norm, not the exception.

Another reason that flying into or out of a large airport may be easier than you think is their facilities. Almost without exception, these airports are very large, have numerous runways available and are equipped with the latest in approaches, lighting and markings.

Probably the biggest exception to the above is Washington's National airport, considered by airline pilots nationwide as one of the most unsafe, scariest airports to fly a large airplane into or out of.

First, Why Do You Want To Fly There?

This first question is not posed to test your courage or ability or to make fun of light airplane drivers. It really is an important question for anyone thinking about operating at these large airports. Do you have a good reason for being there? If, for example, you are planning a business trip to Atlanta in your company's Baron it would be a much smarter decision for you to either fly into Peachtree Dekalb or Fulton County airport. Both are set up for general aviation aircraft, have great FBOs (fixed base operators) and cater to business people. They are both closer to the business centers of the city and very rarely have huge lines of aircraft waiting for takeoff.

If you do have a reason for going to these airports such as dropping off a passenger for the major airlines that operate there, that's fine. Or even just flying there once or twice to prove to yourself that you can is a good reason. But to fly into these huge operations all the time to make some kind of political point about "equal access" is a waste of everyone's time. Sometimes the best way to handle a large airport operation is to choose not to go there at all.

Reasons aside, let's assume you have a need to use one of these monster airfields and examine survivability.

Experience And Preparation

There are two very important points to keep in mind when operating at a large airport. The first is *experience*. The second is *preparation*.

If you have been operating at an airport for a long time, experience is sometimes all you need to handle complex situations when the need arises.

If you are taxiing out at O'Hare on a foggy morning and the Ground Controller says: "Five-one alpha, runway two-seven right, cargo, inner, wedge, outer circular, jog, two-seven parallel, hold short at wolf road," experience would probably be enough to get you to the proper point on the airport if you've been flying out of O'Hare for some time.

To people who are used to operations at that airport the "cutsie" names they have for the taxiways are the easiest thing in the world. To the pilot who doesn't fly there much, they are a nightmare.

The second thing you need for large airports is preparation. Especially if you lack experience there is nothing like having all your ducks in a row before you venture forth into this complicated world.

If you have experience you don't have to do much preparation but if you don't have much experience there is no substitute for lots and lots and lots of preparation.

For the example we gave earlier of a normal taxi-out at O'Hare, instead of experience and a clear mental picture of where the taxiways are, substitute Jeppesen Chart 20-9 for the airport (the airport diagram, complete with the names of the taxiways).

It gives you all the information you need but just having the information isn't enough for an airport like this. If you wait until you get your taxi clearance to look up where you're going you are already about six years behind the rest of the airport. Pull out of the General Aviation ramp at Chicago and block the Cargo taxiway while you look up where the Jog is and you will, no doubt, have a Northwest 747 sitting right behind you leading a line of five or six jets asking on the ground control frequency what you are doing and speculating as to your family background.

Preparing for a busy airport really doesn't take all that much study. In the Chicago example you would have a pretty good idea what runway they were going to send you to by listening to the ATIS (automatic terminal information service). Just before you call for your taxi clearance you could look over the taxi chart and have some general idea of where you would be going. It wouldn't hurt to trace the route ground control gives you with a highlighter on the chart as they give it to you.

Being prepared means having the charts. There is really no excuse for trying to operate at at large airport without the proper charts. How could you handle the taxi clearance at O'Hare without a taxi chart of some kind, either Jeppesen or NOS? It wouldn't be easy, would it? You'd be relegated to asking the controller in a plaintive voice to "give you a progressive" to the runway. Not only is this embarrassing, but it is a huge waste of a busy controller's time and may be in violation of the FARs. Remember that mean, old reg that says that you have to familiarize yourself with all the information pertinent to the flight. It's a large regulation that really means that no matter what you do as pilot in command anything that goes wrong is basically your fault. Many people point to this regulation as proof that "they're out to get us pilots."

We won't argue this point now but why make it easy for them to nail you? Always have the information you need to operate at these large (or for that matter, any) airports. If you don't have the maps and charts, *get them*. If you can't get them borrow them but by all means have them on board before you try finding your way around a huge airport.

At small, VFR airports located out in the country deciding which runway to use is a fairly simple process. You look at the wind sock to decide the wind direction and then pick the runway that is closest to being into the wind.

At large airports they use basically the same system. Usually, whichever runway or runways are in line with the wind are designated as the "active." At many airports, though, they have quite a few other things to take into consideration when they assign the active runway.

Noise abatement is a huge consideration for all airports, but especially those in heavily populated areas.

At Boston's Logan International airport their rules state that if the runways are clear and dry, the wind is less than 15 knots, there is a five knot *tailwind* or less and the weather is VFR they will always assign aircraft takeoffs on runway 15R and landings on runway 33L between midnight and 6 a.m. for noise abatement purposes.

Suppose you are leaving Boston at 1 a.m. and would really like to takeoff into the wind and not with a three knot quartering tailwind...can you demand another runway?

Of course you can. The final authority for runway selection in all cases always rests with you, the pilot in command. It happens all the time and the controllers really don't mind if you request another runway. They *are* required to assign you the noise abatement runway. It is then up to you to request a change.

Many times at places like Boston, airliners have to request another runway for reasons of weight or weather. A Boeing 727 that is going to Dallas, Texas with a full load of passengers and the fuel required for a bad weather day might be too heavy to take off from the noise abatement runway and would request a longer one more into the wind. A general aviation pilot in a Cessna 310 might feel very uncomfortable taking off with a five knot tailwind and would ask for another runway. Either way, the controller will give you another choice assuming there is one. Don't be surprised if you have to take a good sized delay in exchange for the new runway. If they have twenty planes lined up for the active and you are the only one waiting for the other runway you will still have to wait your turn and may have to wait even longer due to other traffic considerations.

Another factor in runway selection for large airports is traffic flow. It is a well-known old saying in Chicago that the controllers launch traffic in the direction they are going and land them in the direction they came from no matter what the weather is. They would deny it, but if you fly there for a while, you *do* tend to notice that kind of pattern.

Speed

If, like most of us, you fly a light single, never forget that most of the airplanes you're sharing a large airport with stall at about your maximum cruise speed. If you're used to operating at a small airport with other aircraft like yours you probably start slowing a few miles from the airport, fly the pattern at 80 knots or so, slowing to 70 on final and 60 over the fence.

Flying that kind of profile at a major airport is a really good way to experience the full wrath of a ticked-off controller. Don't do it.

The best way to get into or out of one of these airports is to fly as fast as you can for as long as you can. The controllers will probably tell you several times to keep your speed up (even if you're flat out), with gentle reminders like the fact that there's a 747 directly behind you.

Plan on flying at cruise or near-cruise speed until almost over the runway threshold, then slow down using a slip. If you have a Cessna (full-flap slips prohibited) don't lower the flaps until you've slowed, if at all. After all, you're not trying for a short-field landing, and you have from 6,000 to 12,000 feet of runway in front of you to play with. Use it. It's okay to float while you bleed off speed—nobody is going to tell you to make the first taxiway if you're unable.

This takes some practice for most GA pilots. We're taught to nail those approach speeds for good reason, but when you're flying into a major airport with a "heavy" behind you it's time to stretch the rules a bit. Practice steadily bleeding off speed while holding altitude, slips, and no-flap landings. If you're fortunate enough to have a large yet uncrowded airport nearby you can practice high-speed approaches.

Once we were flying into Washington National in a Mooney, VFR. The controller gave us the River Visual Runway 18 approach, which takes you down the Potomac from the northwest—a beautiful ride, with a terrific view of the city. The trick is that you have to remain *over* the river, which means maneuvering to follow its twists and turns.

We were told to expect runway 21, and the controller kept us at 1,200 feet (and keep the speed up, please—we were at 120 knots). Just as we crossed the centerline of Runway 18 the controller changed his mind and told us to land on it.

There's always the option to say "unable" and force the controller to do something else, but with almost 7,000 feet of runway available it was not a problem. The throttle came back, the prop went forward, the gear was lowered, and a full-cross sideslip caused us to drop like the proverbial stone. After landing we even got a word of thanks from the controller.

The point of this is that had we done it "by the numbers" we would have had to do a 360 to slow down and lose altitude—not feasible within the confines of the river, not to mention that the controller wouldn't have let us. Refusing to land would have not only earned us a delay while the controller scrambled to fit us into the "heavy" traffic it probably would have gotten him mad at us.

The Major Airports

Let's review the busiest airports you will face in alphabetical order. These are the primary airports at the nation's TCAs.

When the TCA program was instituted, there were only nine Group I TCAs, in Atlanta, Boston, Chicago, Dallas, L.A., Miami, New York, San Francisco, and Washington, D.C. The rest of what are now TCAs were called Group II TCAs. What made a given airport the anchor for a Group I TCA was how busy it was. Those original nine are still the busiest.

Atlanta Hartsfield International Airport

An interesting fact about the Atlanta, Georgia area is that as an urban area, Atlanta has no natural boundaries to its expansion. Other cities like Boston, New York, Seattle, Chicago and even Los Angeles have a large body of water to curb their growth. Even cities like Denver have mountains to block their expansion. It is conceivable, then, that Atlanta could eventually spread out in all directions almost *ad infinitum.*

This isn't true of their airport, Hartsfield International, named after an aviation-minded past mayor. The runways and taxiways of Atlanta Hartsfield have just about reached their maximum size. A few years ago they changed the configuration of the runways and extended runway 9L/27R to a length of almost 12,000 feet, big enough for long-range airliners from all over the world to use.

Atlanta-Hartsfield is one of the original nine airports that had Group I TCAs, and is thus one of the top-ten busiest airports in the country.

Hartsfield is basically a four runway airport with all four of them running east-west. The airline terminal area is located in the center of the field and tends to dominate all the traffic flow on the airport, both general aviation and air carrier. The general aviation area is on the north side of the airport in what used to be the old airline terminal area. Obviously, it is easier for the airliners to get where they are going on the airport than it is for the little guys. Since the airport is oriented towards airline operations this is no big surprise.

The inner runways are almost always used for departures and the outboards are used for landings. The landing aircraft on the outer runways are usually instructed to hold short of the inner ones until

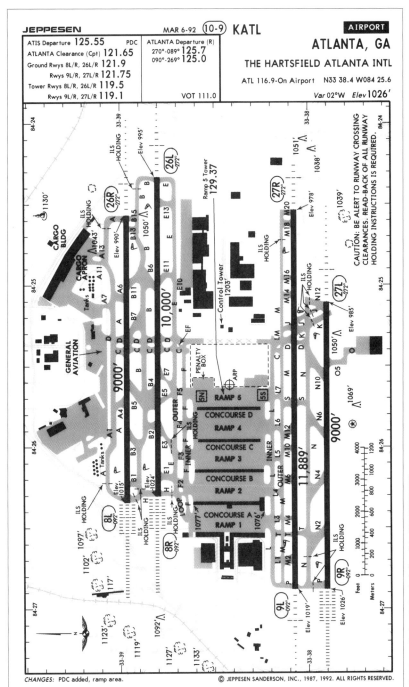

JEPPESEN MAR 6-92 (10-9) **KATL** **AIRPORT**

ATIS Departure 125.55	PDC
ATLANTA Clearance (Cpt) 121.65	ATLANTA Departure (R)
Ground Rwys 8L/R, 26L/R 121.9	270°-089° 125.7
Rwys 9L/R, 27L/R 121.75	090°-269° 125.0
Tower Rwys 8L/R, 26L/R 119.5	
Rwys 9L/R, 27L/R 119.1	VOT 111.0

ATLANTA, GA

THE HARTSFIELD ATLANTA INTL

ATL 116.9-On Airport   N33 38.4 W084 25.6

Var 02°W   *Elev* 1026'

CAUTION: BE ALERT TO RUNWAY CROSSING CLEARANCES. READ-BACK OF ALL RUNWAY HOLDING INSTRUCTIONS IS REQUIRED.

Ramp 5 Tower 129.37

Control Tower 1203'

CHANGES: PDC added, ramp area.

there is a group of them. They are then cleared to cross while the aircraft waiting to take off are held in position.

The airport itself is pretty simple as you can see from the chart...no big surprises. Once you are in the air, either arriving or leaving is pretty straight forward too. Since there are no major bodies of water or mountain ranges around there are departures and arrivals in all four directions, using a "four-post" pattern where arriving traffic is vectored over one of four nav fixes located around the perimeter of the TCA.

The departures from Atlanta are pretty straightforward. There are currently two of them and both involve getting vectored from departure control to a VOR lying on the perimeter.

As with most of the larger airports they expect you to accelerate to either 250 knots or your cruise speed as soon as you can so they can expedite the other traffic. Once again, large airports are set up to handle turbo jets and run their operations accordingly.

IFR arrivals to Atlanta are assigned one of four STARS (standard terminal arrival routes) that correspond to the four directions. No big surprises there either. Except between the hours of midnight and 6 a.m. it is very rare to get any kind of shortened routing into this airport...they live and breathe by these arrival procedures.

Boston's Logan International

Boston's airport is surrounded on three sides by water and on one side by city. Expect all of your arrivals to this airport to be over water. The most commonly used arrival, the Scupp One, puts you thirty seven miles out to the east over the Atlantic ocean...something to keep in mind if you are flying a single engine airplane to this airport, especially in the winter time.

Because the city of Boston is just to the west of the airport the departures in that direction of runway 27 entail a left turn just as you reach the end of the airport.

Logan is another of the original nine Group I TCA airports. There was a much-publicized fracas over landing fees there a couple of years ago, which has now largely subsided—but it's still pretty costly to land there.

As airports go, Boston's Logan is not that hard to find your way around on. There is usually a line for the active runway but if you're a little guy they encourage you to use either the shorter runway or an intersection of the long one to get you out quicker. Once again, it is important and required that you read back all runway and taxiway holding instructions. When they are running parallel operations on

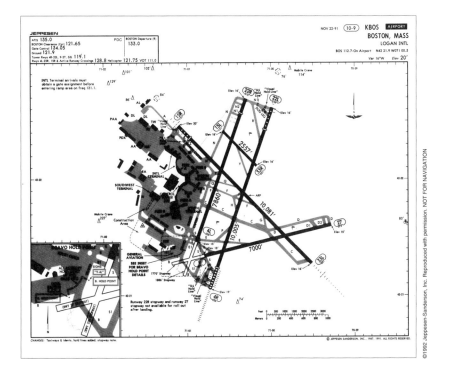

runways 4L and 4R the landing aircraft on the right side usually wait to cross the left for departures just like in Atlanta.

Expect the controllers in Boston to be brisk and businesslike. They, like most busy controllers, have little patience with unprofessional pilots but are eager to help someone who has not been there much and needs help.

Boston is located more or less in a corner of the infamous northeast corridor, and as such has relatively little traffic passing through—most operations in the immediate area are into or out of Logan. Pilots heading north to Maine or New Hampshire are usually sent to the west of Boston, and there is a considerable amount of traffic over Cape Cod and the coastal waters to the south in summer, but not that much humming around the edge of the TCA. This is in sharp contrast to New York, which sits right on top of the most efficient route from New England to the Mid Atlantic states.

Charlotte-Douglas International

Charlotte is a typical large, modern airport. It has two north-south runways, one 10,000 feet long. A third runs diagonally, northeast/

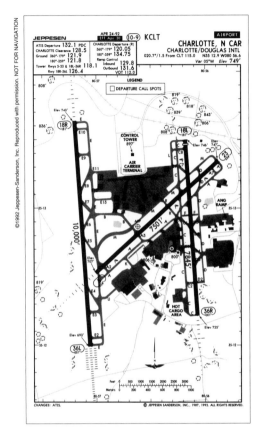

southwest. The north-south runways are the primaries, since simultaneous approaches are authorized. This is important if you're trying to transit the TCA, because it will have an effect on your routing.

Keep in mind where the traffic is coming from and what the controllers are trying to do with it, and plan your route accordingly.

Like other airports that are set up this way, Douglas is run as if it were two airports located right next to one another.

There are two frequencies for almost everything, split along a north-south line: ground control, tower, and departure. There are separate approach control frequencies for traffic below and above 8,000 feet, again split into east/west halves.

Like DFW and Atlanta, Charlotte has a "four-poster" arrival scheme, with an arrival published for the southwest, northwest, northeast and southeast.

There are also two departures, one for jets and one for prop aircraft. These use different routes, nicely separating the big airplanes from the small ones.

Chicago O'Hare International Airport

You've probably guessed by now that this is one of our favorite airports. Located to the west of the city of Chicago, O'Hare was named after a local World War II hero Butch O'Hare.

The VOR identifier, ORD, has mystified pilots for as long as we can remember. The story that makes the most sense to us is that the

JEPPESEN   JUN 19-92   (10-1A)   **Eff Jun 25**   **TCA**

CHICAGO, ILLINOIS

CHICAGO TERMINAL CONTROL AREA

TCA VFR COMMUNICATIONS
(360°-179°) **Chicago App** 119.0   (180°-359°) **Chicago App** 125.7
(VFR transitions N of O'Hare Intl) **Chicago App** 120.55
(VFR transitions S of O'Hare Intl) **Chicago App** 133.5

GREENWOOD/
WONDER LAKE
ILL
Galt

V 217-228

V 24-97
V 100

LAKE
IN THE
HILLS
ILL

CHICAGO/WAUKEGAN ILL
Waukegan
Regl

V 191

D25/ORD

GRAYSLAKE
ILL Campbell

V 24-100-228

D15/ORD

NORTHBROOK
ᴰ113.0 OBK

V 84

V 100-526

100
25

CHICAGO
WHEELING ILL
Pal-Waukee

V 228

100
25

┌CHICAGO O'HARE┐
ᴰ113.9 ORD

D20/ORD

Schaumburg

┌ DU PAGE ┐
ᴰ108.4 DPA

GLENVIEW
NAS ILL

D5/ORD

100
GND

CHICAGO ILL
-O'Hare Intl

D10/ORD

CHICAGO
ILL
Meigs

CHICAGO ILL
Du Page

100
19

D6/ORD

CHICAGO/
AURORA
ILL
Aurora

100
40

100
30

D10.5/ORD

100
36

┌ MIDWAY ┐
ᴰ112.85 XGB

CHICAGO
ILL
-Midway

D25/ORD

GARY IND
Gary Regl

JOLIET
ᴰ112.3 JOT

PLAINFIELD ILL
Clow Intl

┌CHICAGO HEIGHTS┐
ᴰ114.2 CGT

CHICAGO
ILL
Lansing

GRIFFITH
IND

CHICAGO/
ROMEOVILLE ILL
Lewis University

Howell
New Lenox

V 8

V 8-92

JOLIET ILL
Park District

FRANKFORT
ILL

FOR OPERATING RULES AND PILOT AND EQUIPMENT REQUIREMENTS
SEE FAR 91.131, 91.117 AND 91.215

FLIGHT PROCEDURES

IFR Flights-Aircraft within the TCA are required to operate in accordance with current
IFR procedures.

VFR Flights-
a. Arriving aircraft should contact Chicago Approach Control on the specified frequencies.
   Although arriving aircraft may be operating beneath the floor of the TCA on initial
   contact, communications should be established with Approach Control for sequencing
   and spacing purposes.
b. Aircraft departing Chicago O'Hare Intl are requested to advise the ground controller
   the intended altitude and route of flight to depart the TCA.
c. Aircraft not landing/departing the Chicago O'Hare Intl Airport may obtain clearance to
   transit the TCA when traffic conditions permit provided the requirements of FAR 91 are
   met. Due to the traffic density pilots are encouraged not to request such clearance
   during the hours of 0700 to 2300.

CHANGES: Midway VORDME desig.

airport is located on what used to be a small village that was named Orchard Grove...thus the ORD identifier.

In spite of and maybe perhaps because this place is the busiest airport on the planet, you will find that they aren't really bogged down in procedures at this airport. They tend to do what *works,* rather that what's "proper."

At most large airports you will find the controllers assigning jet traffic slower and slower airspeeds as they approach the field and keeping them pretty much on a predetermined arrival route. At O'Hare it isn't unusual to have them tell jet pilots to "fly the 32L localizer inbound and keep your speed up as long as you can." Since the airport is used almost exclusively by jets capable of doing 250 knots the controllers at Chicago have figured out they could handle more of them if they went faster.

Most of the time departing O'Hare follows the same pattern. They usually want you to speed up as soon as you can to get the heck out of their airspace. Once they hand you over to Indianapolis center they slow you down for "sequencing" almost every time. We have no answer for this quandry...i.e.: how come you have to go fast in the busiest airspace in the world and then go slower just when it seems to get less crowded?

We've already discussed the problems you may have taxiing around this airport but the good news is that everything else is pretty straightforward and simple.

Although there are no formal arrivals into O'Hare they follow the same pattern as Atlanta by bringing you in over nav fixes on the four points of the compass.

Keep in mind again if you are flying a single engine airplane that at least one of these commonly used routes takes you quite a few miles out over Lake Michigan—ditching there in the winter time could be hazardous to your health.

This airport like most of the TCA airports has preferred runways due to noise abatement. If you land at O'Hare between 11 p.m. and 7 a.m. expect to use runway 14R. If you are departing during those hours you will be assigned runway 27L.

When you are departing any time of day expect some kind of turn right after takeoff. Sometimes these turns are over 90 degrees of heading change and the controllers want you to do them right now. Of course don't make the turn until you think it is safe to do so but do make that turn as soon as you can. It is a good idea to brief your passengers about this turn because it can be pretty dramatic and more than one passenger has thought they were about to crash when they saw that wing dip.

Cleveland

Hopkins International is the main airport in this TCA. A few smaller, alternate airports you might keep in mind if you go into this are the Burke Lakefront airport and the Cuyahoga airport. Both of these are located to the northeast of Hopkins.

Hopkins holds no real surprises for a pilot as long as you remember where it is located. The field is just south of lake Erie and catches some bizarre weather off the water. It usually snows harder near the lake in winter and you can get some huge thunderstorms in the summer. Another thing to keep in mind about this lake is that ATC will route you over the water as they vector you in from the west or south for runways 23R and 23L. Although you usually don't think of such things, how safe would a single engine airplane be over this water during the fall and winter? You are more than within your rights to request vectors "inside the water" to keep your feet dry in case of an engine problem.

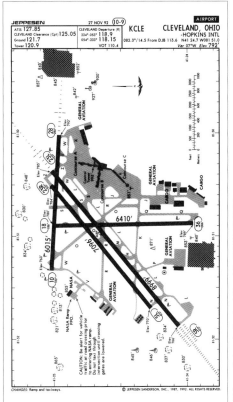

Even though most pilots think of Cleveland as being a flat place without worrisome terrain, you should pay attention to all the radio antennas out to the east of the field. They run from 1,200 feet AGL all the way up to one that is over 2,000 feet tall located within a mile of the final approach fix for the ILS to runway 28! Minimum vectoring altitude to the east of the field is way up to 3,100 feet. If they issue you a visual approach from the east stay high and keep your eyes open.

The general aviation area is to the west of the field right next to the NASA ramp. Expect courteous handling from the controllers and also expect to use the shorter runways, away from the air carrier traffic.

Dallas-Ft Worth International Airport

DFW is a *huge* airport. It has seven runways, countless taxiways and two major airlines that dominate the airport, Delta and American— much like Atlanta. DFW runs its operations off two sets of parallel runways. In Dallas they run north-south but the same pattern as Atlanta, landing on the outboards, taking off on the inboards holds true.

Also like Atlanta the airline terminals are in the center of the airport and the general aviation terminal is off to one side. Unlike Atlanta, DFW has two taxiway bridges that run over the access roads to the terminal. There are four bridges, each one is a one way street and they are numbered by their taxiway numbers, e.g. the "18 bridge." There are two ground control frequencies, one for the east and one for the west side of the airport. You make your frequency changes when you are halfway over the bridge.

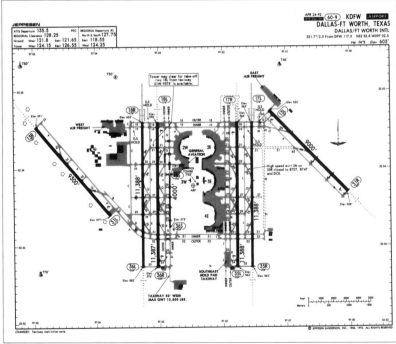

Flying into or out of DFW is difficult during the summer thunderstorm season which seems to run from about April 1 through December 31. Because of the volume of traffic, the approach and departure controllers blank out the weather returns on their scopes. They *are* aware of where most of the really bad weather is and will try to keep you out of it but will occasionally try to get you to fly nearer to a storm than you want. Remember that they are not riding in the seat with you and can't feel the bumps. We've seen people take vectors into weather in this area that looked really bad just so they wouldn't anger the controller. Remember, it's your rear end and your judgment that will come out during the ensuing NTSB trial if you mess up. If it looks really bad *don't* take the vector but tell them why.

Denver

Stapleton International is the biggie airport in this high altitude area. Even though it doesn't look like it if you fly in from the east over the plains, Denver is a high altitude field, sitting 5,300 feet above sea level. It is located on the last flat land before you run into the Rocky Mountains. This location leads to a few problems for pilots.

First, the elevation itself will be a problem if you are flying a normally aspirated airplane and one that isn't pressurized. Enroute altitudes coming in from the east run around 8,000 feet which won't bother you too much, but if you are coming in from or going out to the west, expect to be flying at altitudes of 17,000 feet or above just to clear the terrain.

The terrain to the west leads to another problem for Denver. The winds coming off the mountains can lead to some amazing turbulence at Stapleton. Some of the roughest air we've ever flown through was as we were trying to climb out to the west to go to Salt Lake City. The wind coming over the mountains can make it impossible to climb in even the strongest aircraft. We've been in 727s that couldn't climb heading west during a windy day. Get a good weather brief in this area—it could save your bacon.

Another hazard at Denver is convective activity. Thunderstorms *love* Denver. They've been performing microburst research at Stapleton for a number of years and it was one of the first sites of doppler radar for detection of those nasty phenomena.

The airport itself isn't too much of a challenge, although you might choose a smaller field in the area like Jeffco or Centennial airport to the south of Stapleton. Like all high-altitude airports, long runways are the rule and you could find yourself going on a "taxi safari" to get to the active. Most general aviation activity is limited to the south

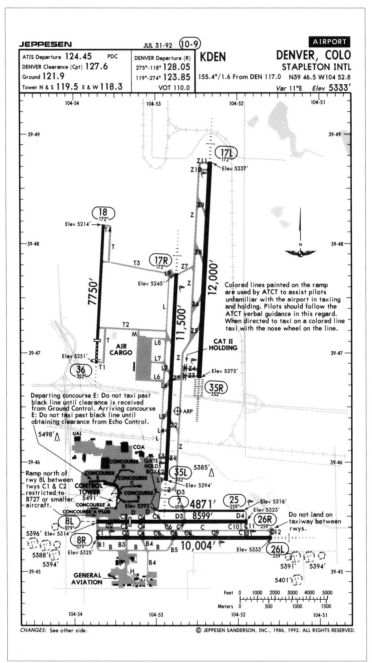

side of the field where you can expect to use the east-west runways right next to the GA ramp.

Denver is a major-league airport. If you plan to go there, do your homework first. There won't be much time to catch up on your reading while you're there.

Detroit

Detroit Metro Wayne County is not one of your happier airports to fly in to or out of. The controllers are grim, the ramp is always crowded and there always seems to be a lot of confusion around the place. We've been flying airplanes there for almost fifteen years and it still isn't easy. You'd be much better off using Willow Run, Grosse Isle or Windsor for your destination.

If you find you have to go to Wayne County, don't panic. It's not that bad if you already know your way around. They seem to have trouble naming taxiways and issuing taxi clearances clearly around there. The controllers always seem hurried and short tempered so be prepared and have a taxi chart handy.

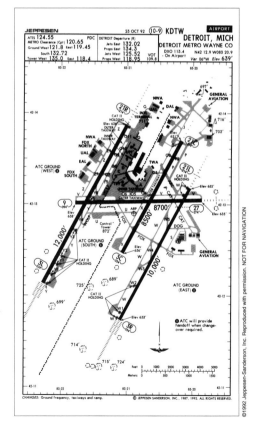

Other than those negatives, Wayne County isn't any more difficult than any other large airport.

Honolulu International

We hope you're fortunate enough to get to fly into this airport. It's located on the southern shore of Oahu, next to Pearl Harbor. An interesting side note is that if you rent an airplane at HNL, the instructor takes you over to Ford Island to do landings

(Ford Island is the one right in the middle of Pearl Harbor, the location of Battleship Row. Shades of Japanese torpedo bombers!)

This airport is unusual in that there's little traffic other than airline activity and commercial sightseeing flights. There is a lot of military traffic in the vicinity, but little of it moves through HNL.

There are two large, intersecting runways and as is common at many major airports the general aviation ramp is on the other side of the field from the terminal.

Staying away from Honolulu and Waikiki beach is tough because of mountains just inland and the broad Pacific offshore.

There's a special flight following service in the Hawaiian Islands that you should take advantage of on any flight there. Called Island Reporting, it consists of making calls to ATC at regular reporting points. Miss a call, and they try to find you. As a result the air-sea rescue record there is impressive.

Kansas City

Kansas City International is really a nice airport but this is one that there is almost no reason to go to because there are nicer, easier to use airports nearby.

International is a huge place out to the west of town, designed mostly for air carrier traffic. Kansas City Downtown airport is the best bet for general aviation types. First, it is located right in the middle of the downtown area which is convenient for business pilots. Second, it has just about everything International does, except the heavy traffic. They have 7,000 foot runways and full ILS approaches.

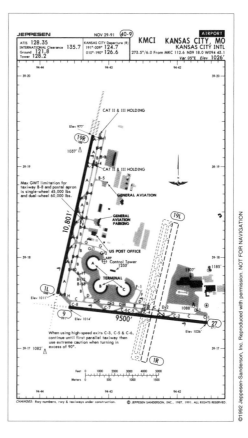

If you do go to Kansas City International keep a close eye on what the controllers are doing to you and other pilots. We say this from experience, but we're willing to figure they've been having a bad day every time we've flown there. Just keep your six covered and pay attention and you should do fine.

Las Vegas

Much like Denver, the terrain clearance cruise altitudes are higher to the west and lower to the east of Las Vegas. McCarran International is the main airport in this area. A good alternate would be Henderson field down to the south.

A big consideration in this area is the military traffic that seems to be everywhere all the time. Nellis AFB is up north and a lot of small, fast traffic comes out of there. Keep your eyes peeled for the military types, especially if you are east of Victor 394 below about 9,000 feet. That's a corridor for them and even though they are looking for you as hard as you are looking for them, remember that they have a parachute and you don't.

Operating in McCarran is a piece of cake. No big surprises at this airport and you are fairly close to the action of the strip and all the casinos.

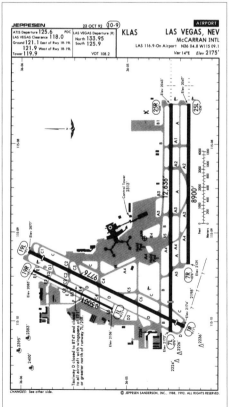

Los Angeles International Airport

For some reason this particular airport seems to have more official arrivals and departures than any other. At any

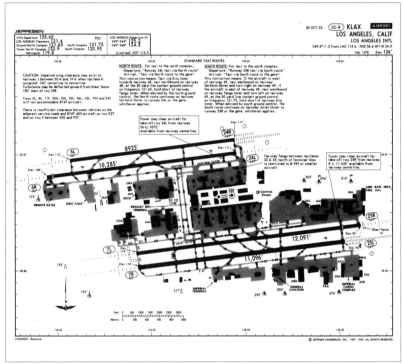

given time there will be about 20 published standard instrument departures and standard terminal arrivals, all for a single airport.

That is the bad news because it necessitates a lot of study and work to successfully arrive or depart from this airport.

The good news is that for the most part, all you have to do is the arrival or departure to work in this area. If, for example you are planning a landing on one of the runway 25s all you would have to do is fly the CIVET THREE (CIVET.CIVET3) arrival.

Everything you would need to make a visual approach to these runways is right there on one chart...no problems.

When departing LAX the procedures help you too because they don't require all that much thought as you are leaving. As long as you follow the departure as cleared you will have no weird clearances to read back as you will find yourself doing on the east coast around New York.

Because, as with all other shoreline airports, the wind usually comes in off the water you can expect to take off towards the Pacific Ocean most of the time. Keep in mind that their noise abatement and

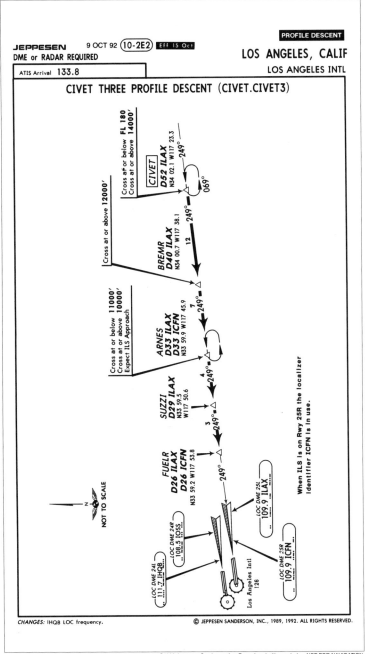

runway use procedures have you flying quite a bit out over the water before you begin your turn.

Don't forget your taxi chart when you go to LAX. It isn't that complicated but the combination of its size and complexity could get you lost pretty quick. Also remember that many of the pilots using this airport only speak English when they are flying and tend to misunderstand the ground controller too.

Jeppesen recently came out with special charts showing preferred VFR routes through the LA TCA. They're a good idea if you operate in this area often.

Memphis

Memphis International is one of the nicer, easier big airports in the country. It is located in fairly flat terrain and the controllers have a firm grip on what's going on.

Professional pilots joke that "you can never get direct Memphis." This is very true because like most big airports, the controllers have favorite fixes to bring you over on your way in. At Memphis expect to come over Gilmore VOR or Holly Springs.

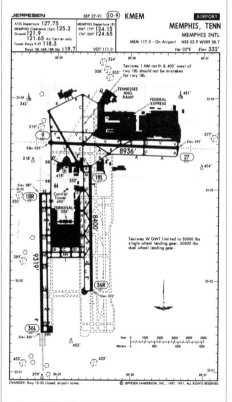

Believe it or not the real busy times at Memphis International are late at night and early in the morning. This is because it's the main base for Federal Express. Expect to see a lot of heavies during these times. The GA ramp is in the middle of the field so every runway is available to you. You will probably get the east-west one to keep you out of the heavy pattern. Memphis isn't a bad place to fly in and out of. Not many air-

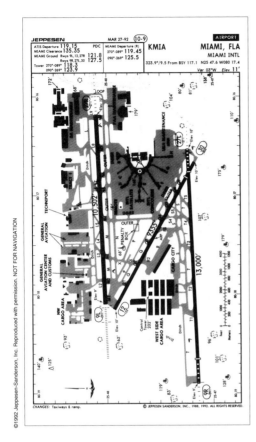

ports have outer markers named "Elvis."

Miami International Airport

Language sometimes is a problem at this airport also. Although everyone speaks English, understanding what is said is sometimes difficult.

The TCA is named after Miami International and this airport is the primary one in this airspace but it is not necessarily the most difficult one to fly out of or for that matter the busiest in the airspace.

Other airports of note in this TCA are Miami's Opa Locka and just on the northeast side of the airspace, Fort Lauderdale, Hollywood International. Further up the beach there is Palm Beach International, Boca Raton and numerous large, very busy airports.

Sometimes, especially when you are operating strictly VFR, the real difficulty lies in making sure you are landing on the airport you think you are. Most of these airports have parallel east-west runways, have the tower and terminal in the middle of the field and they are all located in urban areas near the beach. There has been more than one pilot cleared to land on 9R at Miami International that actually ended up kissing the ground with their mains in Ft. Lauderdale...be careful.

The airport itself presents no huge problems. It is well laid out and easy to find your way around.

In south Florida weather is an ever-present consideration. There is no season when thunderstorms are not present in the area. The controllers are very able and willing to help and with a little care the

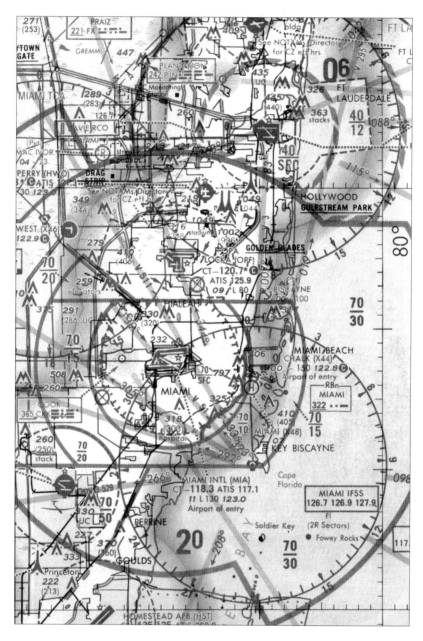

The area near Miami has a large number of airports crowded into a relatively small area. This causes real problems with traffic, particularly during tourist season.

thunderstorms shouldn't present you with any more problems than usual.

Minneapolis

If you don't want to fly your bird into St. Paul Downtown airport or Flying Cloud you will be making an approach into Minneapolis-St. Paul International. This airport shouldn't present any major problems.

Arrivals and departures travel over Nodine, Gopher, Rochester and Redwood Falls VORs. The airport is laid out nicely with parallel east-west runways, an ILS to every one of them and the main terminal placed in between, making it easier for the GA pilot to taxi unhindered to the GA ramp to the south of the runways.

Minneapolis handles snowy weather and snow removal better than any airport we've ever slid to a stop in. Maybe it's because they get so much of it.

KMSP is the home of Northwest Airlines. Be sure to see the huge mural of a 747 they have painted on their maintenance hangar; it's awesome.

New Orleans

New Orleans International is the main 'drome around those parts, but you might want to try Lakefront Airport instead. Lakefront is closer to the city, is less crowded and even looks more interesting than International.

MSY (which stands for Moissant) is the place to go if you like operating around thunderstorms. Expect to see one down there even during the winter and expect it to be a biggie.

Not to worry, though...

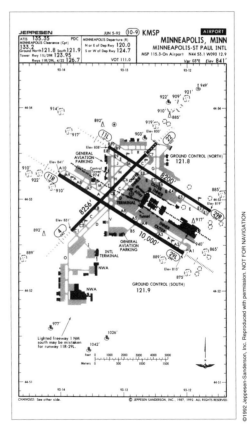

because convective activity is the norm, the controllers are more than used to it and are a lot more relaxed about allowing deviations than the guys at DFW. Ask for a deviation and you shall receive although you can expect to get a little wet no matter what you do.

The main approaches are from the east and west for runways 10 and 28 and there is a back course localizer for runway 19. This approach is over Lake Pontchartrain, but isn't as worrisome as the over water approach at Cleveland, because it is warmer water.

KMSY is one of the sleepier big airports in the US. We've never been there when it was really busy and we've been there a lot.

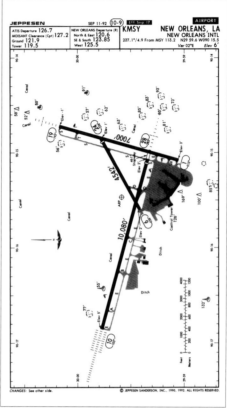

New York TCA

The New York TCA actually has three large primary airports: John F. Kennedy International, La Guardia, and Newark International.

Kennedy International is primarily the domain of international airliners. Although served by domestic carriers, they usually are just there to feed the international routes. This makes for some interesting sightseeing because you get to see aircraft from all over the world, including such rarities as the Concorde.

Aeroflot flies there as do the national carriers of Poland, Ireland, China, Australia and a host of others.

These people are all professionals and operate in a world quite different than we are used to. It might be important to keep in mind if you are landing at Kennedy that some of the other people in traffic

JEPPESEN MAR 27-92 (10-1A) **NEW YORK, NY**

TCA

NEW YORK TERMINAL CONTROL AREA

TCA VFR COMMUNICATIONS

LGA 328°-071° **New York App 126.4** LGA 071°-142° **New York App 125.7**
LGA 142°-231° **New York App 127.4** LGA 231°-270° **New York App 128.55**
LGA 270°-328° **New York App 127.6**
2000' OR BELOW PRIOR TO 8 NM OF KENNEDY INTL **Kennedy Twr 125.25**
2000' OR BELOW PRIOR TO 6 NM OF LA GUARDIA APT **La Guardia Twr 119.95(N), 126.05(S)**
2000' OR BELOW PRIOR TO 6.5 NM OF NEWARK INTL **Newark Twr 127.85**

FOR OPERATING RULES AND PILOT AND EQUIPMENT REQUIREMENTS
SEE FAR 91.131, 91.117 AND 91.215

FLIGHT PROCEDURES

IFR Flights-Aircraft within the TCA are required to operate in accordance with current
IFR procedures.

VFR Flights-

a. Arriving aircraft, or aircraft desiring to transit the TCA should contact Approach
   Control on the frequency depicted for the sector of flight with reference to the La
   Guardia VORDME. Pilots should state, on initial contact, their position, direction of
   flight and destination. If holding of VFR aircraft is required, the holding point will be
   specified by ATC and will be a prominent geographical fix, landmark or VOR radial/s.

b. Aircraft departing primary airports are requested to advise the appropriate clearance
   delivery position prior to taxiing of the intended route of flight and altitude. Aircraft
   departing from other than primary airports should give this information on appropriate
   ATC frequencies.

c. Aircraft desiring to transit the TCA will obtain clearance on an equitable "first-come,
   first-served" basis, providing the requirements of FAR 91 are met.

CHANGES: Grumman Bethpage airport revoked.

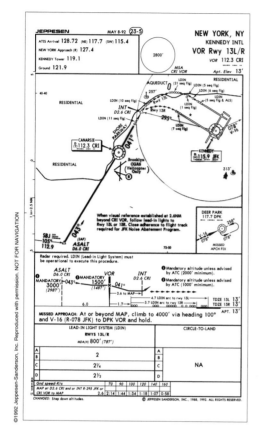

pattern with you may have been in the air for twelve hours or more and are *very* tired. Tired people make mistakes, so watch your six.

Kennedy International has one of the most interesting combined IFR/VFR approaches around next to the River Visual at Washington's National. The approach at Kennedy is called the "Canarsie" approach by pilots although its official name is the VOR Rwy 13L/R approach.

As you can see from the chart it entails flying over the Canarsie VOR and then visually following a series of lights in a curving course to the runway.

The interesting part is finding the correct set of lights because of the combined factors of parallel approaches going onto the other runway and disorienting visual cues from city lights and traffic on a rainy night. You almost have to experience it to believe it.

La Guardia is where the domestic airlines hang out. It is made up of two seven thousand foot runways perched on the edge of the water on the north side of the city. One of these runways—runway 4/22— is built on a pier over the water.

Any pilot will tell you that seven thousand feet is long enough for any kind of airplane to land on...try doing it with a 767 some snowy night. It can get awful short awful fast, especially if you know that if you run off the end you're going for a swim.

They have a nice general aviation area on the south west side of the field and by and large it is an easy place to navigate while you're taxiing.

If you are landing on runway 22 the controllers vector you down the river giving you and your passengers a beautiful view of the statue of Liberty and the city.

Newark International is probably the least regarded of these three airports and is also the most convenient and least crowded for pilots and passengers alike. Nice long runways, good approaches and other good facilities make it a good airport to use if you are planning business in New York—though it's not as easy to get into Manhattan from Newark as it is from one of the other two airports.

The noise abatement procedure is a gentle turn and although it is a busy airport the controllers seem in less of a hurry and calmer than the bedlam you might find at La Guardia.

For an easier time getting into and out of the New York City area we'd suggest using Newark or one of the two relievers for the area—Westchester County to the north or Teterboro to the west, both tucked in under the edge of the TCA. Access to the city isn't as convenient from them as from LaGuardia, however.

There's a VFR corridor with a roof at 1,100 feet that runs down the Hudson River off Manhattan. An excellent sightseeing flight, it can also be a good way of getting through the TCA on days when the controllers are just too busy. Be aware of very heavy helicopter and light aircraft traffic in this "tunnel," and please, keep to the right. The heaviest traffic is found circling the Statue of Liberty.

Orlando

This airport used to be called McCoy AFB before it became Orlando International. B-52s used to be based there and one aircraft remains as a static display on the north end of the field. Because this used to be a SAC (Strategic Air Command) base, it has lots of concrete and very long runways, making it a perfect place for airliners.

International is just to the south of Orlando and is only about seven miles from a much better airport for you if you fly a light GA bird—Orlando Executive. Executive is located in town and is more convenient to things as well as being a very nice field in its own right.

KMCO also is east of Disney World, Sea World and all those other worlds making it one of the most popular tourist airports on the planet. This means that you, the GA pilot not only get to rubber-neck in this area looking for other tourist-pilots, you also have to watch out for airliners taking a little "sightseeing pass." Be careful when operating in this area. Wake turbulence is a real problem with all this heavy traffic so close to the Executive airport and people have died in it.

Orlando International is much prettier than Newark or Detroit

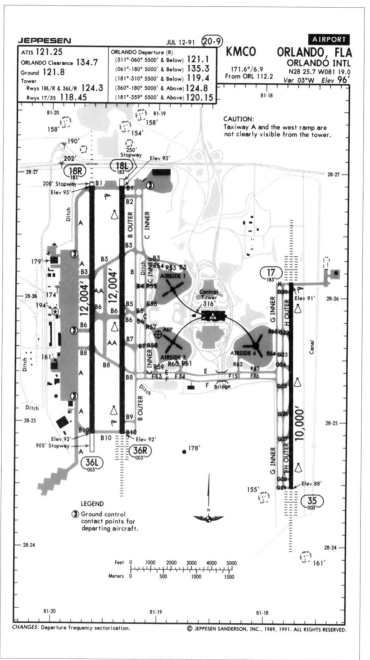

but requires just as much attention to detail and safety as the less glamorous fields.

Philadelphia

International is located to the south of Philly along the Delaware River just across from the navy yard. It's a pussycat of an airport. The runways are laid out well, are long enough and the controllers seem to tolerate the mix of heavy airline traffic, commuter turboprops and GA aircraft without having nervous breakdowns on the air.

Keep the Berlin, New Jersey - Camden airport and Northeast Philadelphia airport in mind as good GA alternates to KPHL. All the airports afford about the same access to the city.

If you are coming in to KPHL by jet expect to be given low altitudes a lot sooner than you'd normally expect. They are trying to get you below the New York arrival and departure traffic and usually won't deviate from that plan no matter how much you complain about your fuel burns. Expect to get a nice, low altitude tour of South Jersey. Other than that, KPHL is just one of your normal, big city humongous airports. Be sure to enjoy looking at all the navy ships in the yard just to the north of the field. You will drag just a few hundred feet over an aircraft carrier if you approach runway 27R.

Phoenix

Sky Harbor International, the main airport, is much busier and more crowded than you might think at first. If you want to avoid the hubbub of mixing with 767s try Falcon field at Mesa or maybe even

Stellar field in Chandler, Arizona.

Sky Harbor is shaped like a shoe box. The two runways are east-west and are over 10,000 feet long. The airline terminal is in between these runways and the GA ramp is located to the northwest.

It is a pretty ride as you approach this airport. Expect to be held high and fast and also count on the controllers to be very busy fast-talking people. All in all this airport isn't that bad but you do have to keep an eye out for diving jets. When they do give these guys lower they really have to hurry down to make the runway. Phoenix usually has windy weather and approaches are almost always an adventure in aviation.

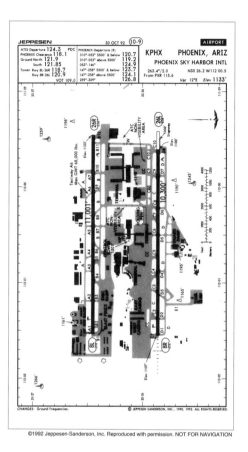

Pittsburgh

This is definitely one place where you should try to avoid the main airport (Greater Pittsburgh International) and use a smaller, outlying field. You have a wide variety of choices of alternates and unless you want to spend the rest of your life taxiing you should avoid KPIT—this place is huge!

The GA ramp is in the middle of a hodge-podge of a sprawled out airport and even with its central location you will probably have a long taxi to get to the active.

The controllers at KPIT are among the best. They are patient, talk slow enough to be understood and have a sense of humor. They need it, because for the first few weeks after a new terminal was put in they had to give progressive taxi instructions to everybody because none of the taxiways were marked and no charts were published!

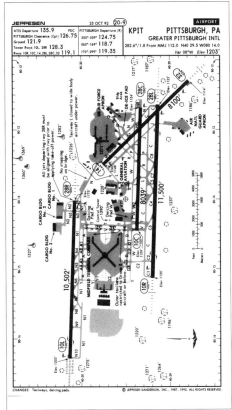

High ground isn't a problem at this airport but keep the high radio towers to the northeast in mind when you get a visual to 28L or 28R. They will definitely get your attention as you cruise by on a semifoggy night.

St. Louis

Once the gateway to the western United States, St. Louis still enjoys an almost perfect location in the center of the country and along the Mississippi river. Lambert - St. Louis International is the main airport in this area and is very accessible to the general aviation pilot as well as the airline types. If you're looking for a slightly smaller airport closer to town try Cahokia airport or Alton, St. Louis Regional airport up to the northeast.

St. Louis is no biggie or challenge for just about any pilot. Its layout is simple, the runways are more than long enough for any aircraft and there is enough room and airspace around the place to give you lots of options when things get tough.

Arrival gates are out to the east at Centralia and Bible Grove VORs. To the west and north you'll travel over Quincy and Foristell as you plot your arrival into KSTL. If you arrive from the east you can give your passengers a great view of the St. Louis Arch as you cross over the river on final for runways 30L and R.

Those parallel approaches to 30L & R as well as 12L & R deserve a little discussion to avoid you having a larger than necessary laundry bill after you fly to STL. The controllers there run parallel visual approaches. They will ask you if you see the traffic they call out to you that will be along side your aircraft as you approach your runway. When this happens, be absolutely sure you see the aircraft they mean. It is real easy to pick the wrong traffic out of the crowd and later scare yourself very badly as you miss the actual traffic by a matter of yards.

The weather patterns in STL aren't very unusual. You will probably get your run of the mill midwest fronts and thunderstorms but there really isn't any big weather danger around KSTL. Keep your eyes peeled for that parallel traffic and operating around St. Louis will be a snap.

Salt Lake City

Salt Lake City International is a seriously high altitude airport. Located in the flatlands between the Rocky Mountains slightly to the east and the Great Salt Lake to the west, this whole area can be a challenge to the experienced mountain pilot as well as the beginner.

Ogden-Hinckley field to the north or the Salt Lake Skypark might be easier alternates but as things go, the big airport shouldn't cause any real problems if you keep in mind where it is.

Its location is the most important thing about this airport because of the weird weather it sometimes conjures up. Winds from the east whipping around the mountains will guarantee you a turbulent ride on blustery days. If it is calm and cool you will find fog around the airport because of its placement even when the rest of the area is clear. About the strangest thing we've ever run into around there was a thunderstorm that was in that valley. Topping out at only 13,000 feet this thing not only was extremely turbulent and full of lightning, it also was a thunderstorm made up entirely of snow—no rain or hail. And it was there in late May, to boot!

Other than the weather the only other thing the small airplane driver has to keep in mind is the field altitude. It is way up there at 4,300 feet and can really cut into your aircraft's performance. Even the airliners have special performance charts for this high altitude field and sometimes carry reduced loads just so they can take off. Also keep in mind that the peaks to the east top out at over 11,000 feet and will definitely cause problems for you if you don't have oxygen and a turbocharged engine.

The controllers are professional and friendly and will do their part to keep you clear of the fighter traffic operating out of Hill AFB which is just slightly to the north.

The view in the Salt Lake area is breathtaking and if you keep an eye on the strange weather KSLC is a piece of cake.

San Diego

The airspace around San Diego's Lindbergh field is a hodge-podge of traffic sectors, VFR corridors, high-speed military traffic and beginner pilot-sightseers. If you don't sit up and pay attention while you're in this airspace you are absolutely nuts. Other than that, Lindbergh field is one of our favorite places on the planet.

Located on the Pacific coast, Lindbergh field is only a few miles away from Miramar Naval Air Station and is within view of the Pacific Fleet sitting at anchor. The airport certainly is in a busy area and has a few other slight problems to boot.

The first small problem is that the airport is located on a flat

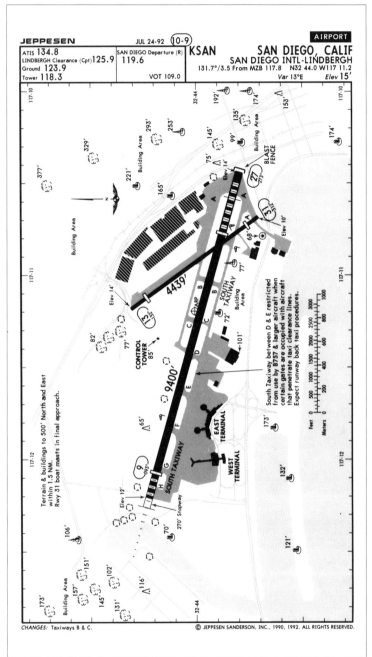

stretch of land with the city up a hill to the east. There is quite a slope up in that direction making a takeoff to the east impossible under certain weather conditions if you are flying a heavy aircraft or one with lousy single engine climb performance.

The second problem, also to the east, is man-made. Some myopic civic planner saw fit to put a huge multi-story parking building just off the approach end of runway 27. This building has changed what used to be an ILS approach with a minimum altitude of 200 AGL to a localizer approach with 700 AGL minimums. If you were flying a 747 on a normal 3 degree descent slope to the runway you would hit the building one floor below the top!

If you can ignore the obvious zoning problem and operate around that man-made mid-air, KSAN is a nice airport. As you roll-out on runway 27 glance over to your right and watch the USMC boots sweat and do push-ups in the sand of the Marine Corps base adjoining the airport.

San Francisco International

Remember San Francisco International Airport? It was the aero-

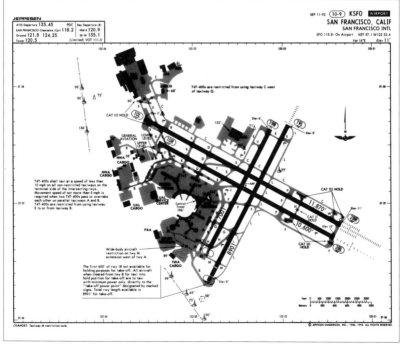

drome that Robert Stack and John Wayne were trying to reach in that epic movie *The High And The Mighty*...the movie that the comedy film *Airplane* was based upon.

The airport is not much like it was described in the movie. It is a very modern, very efficient place that welcomes pilots and aircraft from all over the world. There is quite a mix of domestic and international flights at this field.

As you can see from the taxi chart there are two sets of parallel runways that meet in the center of the field at just about a ninety degree angle.

One thing that is in keeping with its movie image and needs to be remembered, especially by VFR pilots is that there are actually hills to the northwest of the airport that you need to keep in mind if you are departing in that direction. There is terrain above 1,000 ft MSL roughly three and a half miles to the northwest of the field. SFO field elevation is only eleven feet so you can see the obvious problem especially if your bird isn't so hot in the climb department.

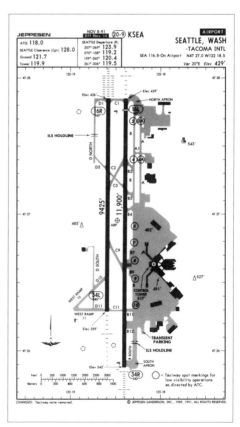

San Francisco is somewhat well known for its foggy weather. The good news is that they have straightforward instrument approaches for you IFR types and for you VFR pilots there are numerous other airports close by that aren't quite so low and close to the water giving you a few safe "outs" if the weather at SFO gets scuzzy.

All in all, San Francisco is a nice airport in a great city handled by some of the most professional controllers in the country.

Seattle

Tacoma International is a very interesting place. Located just to the south of your possible alternate choice, Boeing field. Sea-Tac is a huge place with parallel north-south runways that the controllers run parallel visual approaches to, much like in St. Louis.

The weather? Count on it being foggy or rainy or both. The clear flying days are few in this area but if you are instrument rated or are able to find a clear VFR day you will be treated to great views of Mount Rainier and Mount Washington.

For some reason the controllers in the Seattle area are real nitpickers when it comes to further clearance times, maintaining your position over VFR reporting points and reading back clearances. If you pay attention to detail with these people you should get along.

The Seattle TCA is an elongated north-south affair bounded by the hills and mountains to the east and Puget sound to the west. Expect all the VFR as well as IFR traffic to be funneled in a smaller area than you'd normally expect.

Seattle isn't a terribly busy place but with Boeing field up to the north you do have to watch your six and pay close attention to vectors from the approach controllers around their pattern.

Also, the people around the airport take noise abatement very seriously—right up to the point of litigation, so be sure to follow that noise abatement profile if you're able and 'fess-up if you can't.

Tampa

Tampa International Airport is a very laid-back easy-going place with friendly controllers, a general aviation runway and a light traffic load. If you want an even easier airport in this area try Peter O'Knight airport or the even smaller Vandenberg, both off a little bit to the east. McDill Air Force base is just to the south of KTPA but shouldn't cause you too much trouble.

Two north-south runways having full ILS approaches make up the main actives at KTPA. The general aviation runway mentioned above is an east-west that is almost 7,000 feet long. Taxiing around this airport is easy and the general aviation area is well separated from the airline crowd and is located to the south of the field.

Once you are airborne expect ATC to get a little snippier if you head east toward Orlando and the Disney World area. These controllers not only have to contend with huge amounts of airline and military traffic they also have to deal with thousands of sightseeing

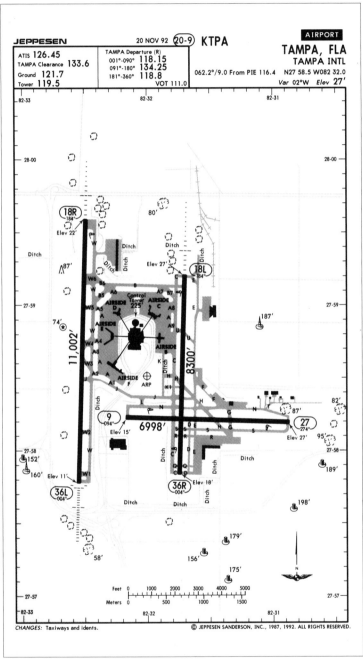

light plane drivers that are in a partying mood and might not be paying much attention to their traffic advice. You shouldn't become one of them.

Anywhere in Florida, but especially in this area you should have your eyes out of the cockpit looking for banner towers, sky diver planes and hot air balloons as well as your basic tourist in a Cherokee.

Washington National Airport

Earlier we said that this airport was considered by many air transport pilots to be one of the scariest and hairiest aerodromes in the United States. That is only partially true. There are some pilots that will claim that it is by far the absolute scariest airport in the country.

Why? The airport really isn't so bad. It is, as you can see from the chart, made up of three semi-decent runways, the longest being almost seven thousand feet. It has a simple taxiway system and seems easy enough to understand how it is laid out.

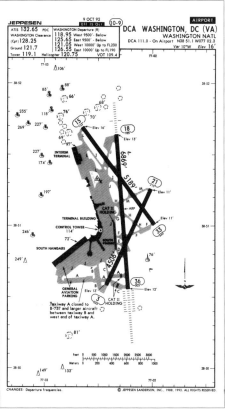

Now take a step further back and notice where it is and what where it is, means. (See the chart on the next page.)

The airport is surrounded on three sides by the Potomac river. To the north of the airport you will notice not a restricted area, not a warning area, no, you will see a **Prohibited Area.**

This Prohibited area is less than three miles from the end of runway 36. This is not a piece of airspace

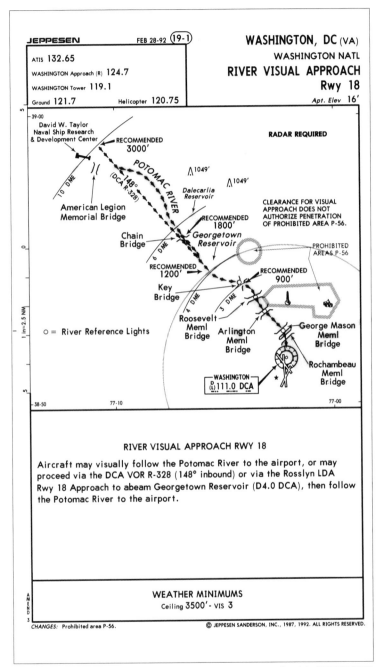

that will just get you a nasty letter and a slap on the wrist if you fly into it. It can also get you shot down. Remember during the Vietnam war when a disgruntled person tried to fly a Huey into Richard Nixon's oval office and say "howdy?" According to various sources, after that incident the Secret Service has *at least* shoulder-fired SAMS at the site...maybe worse. Feeling lucky today?

Actually you would be lucky if you lost an engine on a twin after taking off on runway 36 and made it to the White House to be shot down by people in dark glasses. You would first have to miss the Washington Monument, a small six hundred foot high airplane grabber.

What if the wind is from the south? That would make your departure much easier...how about your approach? Check out the River Visual Approach. Looks simple until you realize that as you do this procedure of yanking and banking, you are squeezed between the river and the prohibited areas on your left.

Something they don't show on this chart is there are buildings out there. We're not talking about two story houses...we're talking high-rise office buildings all along this river right up to where you make your dramatic right turn about half a mile from the impact point on the runway.

Not complicated enough for you yet? Try doing it three miles behind a 727...at night...in the rain...or snow. All they need is better than a ceiling of 3,500 ft and more than three miles viz and they run this thing.

Okay, big deal. You land a little long because this was one hairy-muffin of an approach. You start to slide on the runway because it is a little wet or icy and it looks like you'll accidentally run off of the end a little bit. No problem at most major airports other than the fact that you'll have a lot of explaining to do to the FAA. At this one, you're going for a swim. Right after the Air Florida disaster they had a real shortage of rescue boats to pick up the people...how many do they have now?

We're sure Washington National Airport has a pretty good safety record if you look at the statistics. We think it is because pilots are so pumped-up worrying about whether or not they will safely make the approach or departure that they pay more attention to things when they operate there.

It is embarrassing that we, the most powerful nation on earth have this kind of airport at their capitol city. When you fly into or out of National watch your six.

Planning and Departure

At the beginning of this book, we took you on a fictitious flight from Burlington, Vermont to New York's LaGuardia with a private pilot in a Cessna 182. The next three chapters will go over that flight again in more detail, to give you a close look at some good ways to safely and efficiently work your way through crowded airspace.

The pilot in our example was not instrument rated, but all of the techniques we discuss here can be used just as effectively by IFR pilots as by VFR pilots.

Our example assumed VFR conditions so our pilot could make the flight. Had he been IFR rated, he would have had the luxury of choosing whether or not to file, even though the weather did not force him to. Before we examine the trip from Burlington to New York, let's look at the choice between VFR and IFR in more detail.

VFR or IFR?

Of course, the decision of whether or not to file is driven mostly by the weather. But, if we assume it's VFR along the route other factors come into play.

In many ways, flying IFR is easier than flying VFR. The entire ATC system is oriented towards the instrument pilot, with VFR pilots being accommodated on a workload-permitting basis. You have the benefit of a well-defined route to follow, constant monitoring of your flight should something go wrong, separation from other IFR traffic, and in many parts of the country radar coverage for the entire flight.

On the other hand, when flying on an IFR clearance you have to take what ATC hands you even if it's not the route you wanted. Also, sometimes you must deal with hassled, stressed controllers who are talking so fast it's a wonder they can catch their breath when a VFR pilot could just cruise along unperturbed.

It's our preference to fly IFR when possible, even on VFR days. But, we don't always do it. There are four main reasons to fly VFR even if you're instrument rated.

First, there's routing. After flying in a certain area for a while it's possible to anticipate where ATC is going to send you, no matter what you ask for. Often it's far enough out of your way that filing simply isn't worth it. Why add 50 miles to the trip if you don't have to?

The second is time. Even when the weather is VFR, an instrument pilot should always take the time to properly set up the route, gather plates and charts, and so forth. Plus it's necessary to file a flight plan, get it into the system, and wait for a clearance. All this takes time, particularly waiting for the clearance. There are times when the weather is fine and you simply don't have time to wait for ATC to come up with a clearance for you.

The third is when the flight is so short it doesn't warrant filing. That includes operations in the local area. There's really little point if filing IFR for a 20-minute hop from one airport to another, or if practicing approaches.

The last is special considerations. In particular is the IFR slot reservation system in use at some large metropolitan airports (JFK, LaGuardia, Washington National, and O'Hare). When flying IFR into any of these, it's necessary to call in advance (no more than 48 hours) and get a reservation for an arrival time. You are required to actually arrive within 15 minutes of this time in order to get in. Often, it seems that the times are always full when you want to arrive, and you're forced to make big changes to your plans just to get there when they'll let you in. The nice loophole here is that a reservation is not needed for VFR pilots. You just show up, call approach before entering the TCA, and they'll let you in no matter what time it is—that is, unless you're flying to LaGuardia. At that airport and no other you can't get in VFR at certain times of the day, period (more on this later on).

What it all comes down to is expediency. If the object is to get you where you want to go, you should choose the means by which you can accomplish that most effectively.

There's another alternative, however—the combined flight plan. In the chapter on small airports we described Westchester County,

New York and its surroundings. It's tailor-made for combined flight plans.

When flying to the southwest, the usual routing is to be sent west of White Plains to Sparta VOR, then turned south. This often adds considerably to a trip's length. Better would be to go south to Colt's Neck VOR near the New Jersey coast and proceed from there. A good way to guarantee that you can do that is to file so that your route starts over the VOR; you depart VFR, fly down the Hudson River corridor (never entering the TCA) and when clear call ATC for your clearance.

Let's assume you've decided to go VFR, and examine the process.

Preplanning

In our sample flight the pilot ("John Smith") made the decision to pick up a client at LaGuardia for a flight to Glens Falls, New York. He began the process a few days in advance, not a bad idea for a big trip to an unfamiliar area.

The process starts with gathering information. The pilot bought current sectional charts, a VFR terminal chart for New York, a current A/FD, enroute charts and a book of approach plates covering New York.

He also looked at a commercial airport guide to get the phone number and location on the field of LaGuardia's sole FBO.

Why all this information? A glance in the Airport/Facility Directory for New York provides an excellent example of the value of doing your homework. There's a page outlining the IFR reservation procedures for the four high density traffic airports in the U.S. (Kennedy, LaGuardia, Washington National and O'Hare). At the bottom of the page is this note:

"VFR reservations via ATC for non-scheduled operations are not authorized between the hours of 0700 to 0900 local and 1600 local to 1900 local, Monday through Friday and Sunday evenings unless otherwise announced by notam."

In other words, if Smith arrived north of the TCA between 7 a.m. and 9 a.m. and told approach control he wanted to land at LaGuardia—exactly what he was originally intending to do—they would have told him to go away and come back later. We've looked elsewhere for this tidbit of information, and this deftly-buried passage is the only place we've ever seen it. It's not on any chart or approach plate that we've found.

That evening, he called the FBO to find out about landing fees, parking, a waiting area for his client, fuel availability and costs.

He also went over the charts, starting with the New York terminal chart to plan his arrival into the area of the TCA.

If you're flying into a major terminal area, this is an important step. It will show you what to expect so that you don't arrive "cold."

From this chart, you should note the name and frequency of the nearest VOR to the edge of the TCA (Carmel, in this case), along with the proper frequency for the initial callup of approach control.

The reason for this is to provide you with a known, prominent point to fix your position for the controller when you call, one that you won't have to worry about identifying visually. It's a safe bet that when flying into a Class B/TCA you'll be given radar vectors to follow, so detailed route planning inside the airspace is not needed. Your planned route can stop at the final VOR fix. If flying into an ARSA, however, route selection is up to you.

Smith also looked over the approach plates for LaGuardia to see where traffic would be funneled in from for the various primary approaches to each runway. It's unlikely a controller will try to mix VFR traffic with jetliners until the last possible moment, so the approach routes are places you'll probably be steered clear of. Which one is in use depends, of course, on the winds.

Next, the enroute and sectional charts came out to choose a route that would begin at Burlington and end at Carmel VOR. Here, there are several factors to weigh, including ease of navigation, directness, and safety. It's better to fly a route that takes you over several airports at the expense of a few miles than to take a direct shot over a mountain range.

In this case, that's what Smith decided to do. Victor 487 heads south from Burlington directly for New York, but it turns eastward around the Connecticut/Massachusetts border. Smith would be navigating direct from VOR to VOR from there on. Also, he'd be over rugged terrain for much of the flight, and would only pass near one airport in Great Barrington, Mass. Instead, he decided to use Victor 91, which splits off of Victor 487 and heads over to Glens Falls and Albany, where he'd pick up Victor 123. That would take him straight to Carmel VOR, keep him over wide river valleys, and take him over three large airports along the way. The difference in distance was only about 10 miles.

The only remaining thing was to choose an alternate in case, for some reason, he couldn't get to LaGuardia. The obvious choice was Westchester County, lying between Carmel and LaGuardia. He located it in the approach plate book, and marked it with a paper clip. He did the same for the three airports he'd be overflying (Glens Falls, Albany, and Poughkeepsie). That way, if he needed to contact them all he'd have to do is flip the book open.

The last preplanning step was to compile all the information gathered so far. The route was marked on the IFR chart, and all the frequencies he'd need were written down, including ATIS, clearance delivery, ground, tower and approach control frequencies for Burlington and LaGuardia, and all of the VORs enroute.

The Weather Briefing

The night before the trip, he used his computer to call DUAT for a quick look at the synopsis and notams. This gave him a preliminary picture that would be accurate enough to tell him whether or not to cancel the flight.

The synopsis told of a cold front that was expected to pass through the area before dawn, leaving turbulence and strong winds behind it.

Before dawn on the day of the trip, Smith got up, picked up another full DUAT briefing which he printed out, collected his prepared materials, and flipped on the television.

He spent a few minutes watching *The Weather Channel*, long enough to get a look at a satellite photo and weather map. He also figured out his time en route from the DUAT information. A look at the printout showed him he wouldn't need to change the route due to weather.

Armed with this, he called Flight Service and got a live briefing. Smith liked to read along in the printout as the briefer talked—it let him absorb the weather better.

Alternative Weather Sources

This is still the most common form of briefing, but the alternate methods are gaining popularity. Many pilots like to get a full briefing from DUAT and carry it with them. This eliminates the possibility of missing something the briefer said, but a complete briefing can take up many pages and can be daunting to wade through. The DUAT and other computer weather providers (like Jeppesen) also offer custom-tailored charts, but these cost extra. The basic service is free, and what you get is the same raw data that a Flight Service briefer sees.

Another excellent source of information are the fax services. There are several of these, and their product is excellent—the same official NOAA weather charts that are posted at the Flight Service Station, itself. The drawback is cost. Not only do you need a fax machine or computer with fax modem, you have to pay roughly $1.50-$2.00 per sheet for the charts.

The "old-fashioned" telephone briefing still has its benefits. There's a live person on the other end of the line who can offer some limited help in interpreting the raw numbers for those who aren't ace meteorologists, but a pilot should never forget that it's not the briefer's function to make decisions for him. In fact, it's possible to a briefer to unintentionally put a "spin" on the weather that leads a pilot to believe he can make it when he really can't. (We know of an accident in which the weather was questionable, and the VFR pilot called for three weather briefings. Each briefer read the facts, but one sounded a little more optimistic than the other two. The pilot took off after that briefing, and didn't make it.) The bottom line is that the pilot is in charge and must make his or her own decisions.

When you call Flight Service, the kind of and quality of the briefing you receive will depend largely on you. The FSS briefers are of the highest quality. They are trained to give you the best information possible and are usually experienced pilots in their own right. They really want to give you valuable aid for your flight planning. They only restriction they face is you.

When you call, or even visit in person it is you that will dictate the kind of briefing you will get. If Smith had asked only for the Burlington and New York weather that's all he would have gotten. Although the briefers will try to prompt you to ask for the information you will need it really is your job to know what you need in terms of preflight information.

How much information is enough, then? The FAA thinks it is best for you to provide at least eight pieces of information to the briefer in order to get the best information. Take a few seconds at the onset of the briefing to provide the briefer with the following information.

• **Type Of Flight Planned:** VFR or IFR. If you're instrument rated and the weather seems good just tell the briefer that you are both VFR and IFR qualified and are trying to make up your mind which way to go today.

• **Aircraft Number:** Before you even begin it is important to identify yourself as a pilot first of all...believe it or not, many people call the FSS to figure out whether or not to mow their lawns. Usually the briefer will ask you for an "N" number or aircraft number to identify you as a pilot before they go any further with the briefing. It can also help to tell the briefer your rating—if you say you're a student pilot you won't waste a lot of time finding out that the weather is beyond your qualifications.

• **Aircraft Type:** This is an obvious one. If you are flying a Gulfstream III today your briefing would be much different than if you are flying a 172. You might want to include any special equipment you have at this point also. If you have an inertial nav or RNAV capability it would be nice for the briefer to know.

• **Departure Airport:** In this case, Burlington International Airport. The exact airport is important because part of the briefing will be notams for that particular airport.

• **Route Of Flight:** Just give an approximation of the way you think you are going for this one. Obviously your final route will be predicated at least partially on the weather the briefer is about to tell you about. In our example, something like "via Albany" is sufficient.

• **Destination:** In this case, LaGuardia, New York.

• **Flight Altitude:** There is no need for the briefer to give you forecast winds aloft for flight level 280 if you plan to cruise up there at ten thousand. This is also extraneous if you're flying a non-turbocharged airplane. In that case, the only winds you could possibly use are 3,000, 6,000, 9,000 and 12,000.

• **ETD:** Estimated time of departure and time enroute are important for the briefer so he or she can give you weather that means something. If you aren't leaving for five hours and just want a background briefing it would be pointless and a waste of time to give you current conditions. A rough idea of time enroute (say, to within an hour) will also prompt him to give you the appropriate forecast.

Once you give the briefer this information, and remember that it should only take you a few seconds to run down all of the above, he or she will try to give you the "big picture." They will try to communicate a "picture" or summary of the weather conditions in the area and what you can expect on your flight. They won't read weather reports verbatim unless you request it, but will include the information in a standard format.

A Standard Briefing

If you haven't received any other information or briefings for this flight you would probably request a "standard briefing," also sometimes called a "full briefing." The briefer then has nine mandatory things to review for you. They are usually given in the order in which we will review them and it only takes a few minutes for them to do this. That is why it is a good idea, both for remaining legal and safe...this briefing format should provide you with the information you will need.

• **Adverse Conditions:** These are items that might convince you not to fly today...at least to that particular destination. These include low ceilings, poor visibility, icing, convective activity (thunderstorms), precipitation, turbulence, Airmets, Sigmets, and so on.

• **VFR Flight Not Recommended:** If you're a VFR pilot and hear this, *pay attention!* If you tell the briefer in your initial call-up that you propose to fly VFR and the conditions aren't favorable for flying visually they will tell you at this point. They are in no way trying to make your decision for you. Whether or not you press on VFR is up to you. Just remember that if you prang it after getting a "VFR not recommended" briefing it may not sound too good at the hearing—assuming you're there for it.

• **Synopsis:** This is a brief statement describing the type, location and movement of weather systems that might affect your flight, and what they're expected to do for the next several hours. If you've seen a weather map in the last few hours (on television, for example) you'll already know about the systems the briefer is describing and will be able to visualize them easily.

• **Current Conditions:** If you are within two hours of your proposed departure time the briefer will provide you with current conditions at both your departure and destination, PIREPS and other information.

• **En Route Forecast:** The weather forecasts will be reviewed in the same order as your proposed flight; i.e., departure-climbout, en route and descent.

• **Destination Forecast:** The weather forecast for your destination airport from one hour before your ETA to one hour after.

• **Winds Aloft:** The briefer will ask you for the altitudes you want, and will usually provide forecasts for a location near your departure point, two or more locations along the route, and one near your destination. Most GA pilots will be interested in winds at three, six, nine, and 12,000 feet. This will give you the chance to see what the most efficient altitude will be. This information is presented fairly quickly, so it's a good idea to develop an organized method for copying it fast.

• **Notices To Airmen (Notams):** The briefer will review notams for both your departure and arrival airport as well as those on your proposed route of flight. They are also required to review those within a four hundred mile radius of their FSS. The notams a briefer reads are called "D" (for distant) notams. Some notams are eventually published on charts and approach plates, at which time they're dropped off line and the briefer doesn't see them. Also, some notams are of a local nature ("L" notams), things like "caution for deer on the runway" or snow removal advisories at small airstrips. The airport manager is responsible for filing these, and sometimes he or she doesn't get around to it...so they're a bit suspect. A briefer won't look these up unless specifically asked. Most of the time it's unnecessary, but if you're flying into a very small, sleepy airport it might be worth your while to request them. It wouldn't do to arrive at Podunk Regional and discover that the runway is closed because the snow-plow wouldn't start that morning.

That ends the required part of the FSS's briefing. You may request any other information from them that you feel is important to your flight.

There are three other kinds of briefings you can obtain from an FSS that might help you out.

If you have already been briefed or just need a few pieces of information you can request an "abbreviated briefing". Just tell them what you want and when you had your last briefing and they can fill you in with just the information you need.

An "outlook briefing" is one you would request if it was going to be six hours or more before you plan to leave. This would be a sort of background briefing you might get before you go to bed the night before a proposed flight.

You can always get an "in-flight briefing" anytime you want over standard FSS frequencies or Flight Watch on 122.0.

With all these briefings the FAA hopes you will wait until the end to ask any questions. They will probably cover your question somewhere along the way anyway and this will save time. If you do have any question after the formal briefing be sure to ask it. Remember, according to the FARs it is your responsibility to know *everything* pertinent to your flight.

Review the Information

Many pilots forge right ahead at this point and file a flight plan. We prefer to review the information, see how it affects our planned route, make any necessary changes, and come up with a "hard" ETE estimate based on the winds.

Smith had already done this, using his DUAT printout. He filed his flight plan after the briefing was over. Over breakfast he reviewed the route one more time, did some calculations and filled out his flight plan. He also sketched in a rough flight log. with headings and frequencies from the list he'd already prepared.

File A Flight Plan

Once you've received a good weather briefing it is always a good idea to file a flight plan, even if you are VFR. Obviously, if you're IFR you have to file to get a clearance. It would take another chapter to tell you all the good reasons why you should file a VFR flight plan, but we'll just give you two here:

1. It's free. It doesn't cost you a penny to get the extra protection a VFR flight plan affords.

2. If you're flying out of a busy airport (like Burlington) you'll need a clearance to leave the airspace anyway. Why not take the extra minute after your briefing to file?

Departure control won't open your VFR flight plan for you, they're usually too busy. Just wait until you're out of their coverage, slip over to the FSS frequency and open it then. Just remember to open it on your time of departure, not your time of call-up so it will be accurate. On the other end, cancel it before you contact approach control. Once you're in a busy terminal area you're being monitored constantly, so the protection of a flight plan is no longer pertinent, and canceling early means you won't have to remember to after arrival.

Out On The Flight Line

After Smith arrived at the airport and preflighted the airplane, he climbed in and prepared the cockpit. This is a good habit, and should be a definite part of preflight operations.

First, all of the necessary charts were pulled out of the flight bag. That included Montreal and New York sectionals, the L-25/26 Low Altitude Enroute Chart (which covered the entire route), and the New York VFR terminal chart. Also, the book of approach plates for New York and his clipboard with flight log, frequency list and weather information were set out. Everything else went in the back seat. The charts were opened to the appropriate panel and stowed in the map pocket, a rubber band was placed in the book at LaGuardia's plates and the book stowed, and the clipboard prepared.

A Few Words About ATIS

The ATIS (Automatic Terminal Information Service) is simply a means by which the controller is relieved of having to read the same information to every pilot that calls. It saves time and keeps the frequency clear. Only towered airports have ATIS, and it's only updated when the tower is open. An increasing number of small airports have AWOS (Automatic Weather Observation Service), a kind of pseudo-ATIS that employs a robot weather station and synthesized voice to issue weather conditions.

The information in the ATIS contains (at least) the ATIS code, which runs from A to Z and is changed every time the broadcast changes, the UTC time that the broadcast started, the weather, and the runway and approach in use. In addition, any local notams or other information will be tossed in. The ATIS changes every hour or more often if something else happens that warrants a new broadcast.

The ATIS at a major airport can get very complex. Here's a typical example for Atlanta Hartsfield:

"Atlanta Hartsfield International information Golf. All aircraft shall read back all runway holding instructions including aircraft identification. Pilots are requested to read back their transponder code only unless company policy requires a full read back or they have a question. The 1500 UTC weather: five hundred scattered two thousand five hundred overcast, visibility one and one half in fog. Temperature five zero, dew point four nine. Wind three four zero at one eight, gusting to two six. Altimeter two nine nine zero. Departure runways two six left and two seven right. Runway two six right closed. All aircraft contact clearance delivery on one two one point six five prior to taxi. Gate hold procedures are in effect. Advise the controller on initial contact you have received information Golf."

Quite a mouthful. You should listen to the ATIS as many times as it takes to get it all—there's no hurry. You should also write it down, no matter how simple the information is.

The example above is pretty straightforward, though lengthy. Just about the only thing that you may not have heard before is the term "gate hold." This is an airline term used to describe a delay that they used to take in the air in the form of a hold.

Let's say that Delta has a flight going from Atlanta to Chicago at around four p.m. Let's also say that bad weather has really slowed down O'Hare's operation. In the "good old days" of airline flying (prior to the ATC strike) the Delta flight would take on extra fuel, fly up to the Chicago area and be put over a holding fix, sometimes for hours, until ATC could sort things out and allow them in the line for the approach.

A gate hold allows the Delta flight to take its delay on the ground in Atlanta. The flight would still be in the same spot in line for approach in Chicago, they would just be doing their holding on the ground. No fuel would be unnecessarily burned, the flight would arrive at the same time it would if it had held in the air and safety would be served because there would now be a limited time the flight would be in the air operating around the nasty weather.

The gate hold delay is also called having a "wheels-up time." In the case of your flight, especially if you were going to another busy airport, your ground delay would be basically the same thing as the airliner's. You would be held on the ground until ATC was ready for you to enter their system.

"Flow control" is another term you'll hear if you travel in busy circles much. A flow control delay is one usually involving the ARTCC Center's airspace. They have a maximum amount of traffic they can safely handle and will limit the amount of traffic when they have to. Many times you'll run into a flow control situation when either the Center has a problem like their radar or computer being down or if there is a weather problem like a line of thunderstorms in their airspace.

If you operate IFR, all of these delays might be a factor in your flight. If you leave Atlanta VFR only local traffic congestion (pushes) would cause you problems.

If you're flying VFR out of a plain-vanilla controlled field (i.e. one outside a Class B/TCA or Class C/ARSA), you can fire up your engine and listen to the ATIS while it warms up. But if you're IFR on a nasty day, or operating out of a big airport anytime, it's wiser to listen to the ATIS before you fire the engine. In between the ATIS and waiting for a clearance you may be on the ground for quite a while, and there's no good reason to waste fuel while you wait for ATC to let you take off.

Get The Clearance

This step only applies to IFR pilots or those departing from a Class C/ARSA or Class B/TCA airport.

Most medium to large airports today have a discrete frequency for clearance delivery. At smaller airports the clearance is obtained from ground control. At some uncontrolled fields there's even a remote communications outlet to a nearby towered airport, so you can get your clearance over the radio even though there's no tower.

In the case of Mr. Smith, who is flying VFR to LaGuardia, his next step is to call Burlington's ground controller and get a clearance out of the ARSA. The wording goes something like this:

"Burlington ground, Cessna 1234 at the general aviation ramp with Sierra, VFR to LaGuardia, New York."

The destination is necessary because your frequency assignment may depend on the direction you're headed.

The controller will respond something like this:

"Cessna 1234, Burlington, after departure fly runway heading, maintain 2,000, contact departure on 121.1, squawk 5342, advise ready to taxi."

This should be written down as the controller says it, and read back. Since you haven't started the engine, an addition like "we'll be ready to taxi in five minutes" is helpful if the airport's busy to let the controller know he need not pay attention to you for a few minutes.

If you were IFR the initial call to ground or clearance delivery would be the same only saying that you're IFR. In this case the controller is likely to say:

"Cessna 1234, Burlington, clearance on request".

This doesn't mean that you're supposed to request your clearance. It means that they've passed the word to the computers that you're ready to go and want your clearance.

Depending on the weather and your route, you may be waiting for quite a while. We've had to wait as much as half an hour for a clearance. This is the number-one reason for not starting your engine until actually ready to taxi.

You'll need to sit there as long as it takes, listening to the radio and waiting for the controller to say:

"Cessna 1234, Burlington, advise ready to copy."

When you say "ready," you should *be ready*. You're about to get a long stream of information read quickly, and if your pencil isn't poised you'll miss some of it.

If you're lucky you'll be cleared "as filed" and after readback you'll be ready to go. Often you'll get a full route clearance, which means that ATC didn't give you the route you asked for. If this is the case, take a moment right now and go over the route on your chart to make sure you know where you're going. We've had clearances read to us that were plain wrong, sending us to intersections that weren't even on the airway we were supposed to be flying on. A highlighter is handy in this situation, since it lets you mark out your route quickly.

If at a large airport there may be gate-hold procedures in effect. If that were the case you'd get something at the end of the clearance like "Expect wheels-up at 1755 zulu," or they would ask you to contact a discrete flow control frequency for a time slot before you started your engines. In a situation where there is less than thirty minutes of delay expected you should probably start-up and get in line for takeoff—these procedures are only put into effect at large airports where it's very common to have long lines of airliners waiting to take off. Sometimes if you get there early they can let you go five minutes sooner than they told you. At any rate, it will probably take you at least twenty minutes to work your way to the front of the takeoff line anyway. If you are approaching the end of the runway well ahead of your time ground control will probably have you wait on an intersecting taxiway.

If you are too late getting to the end of the end of the runway and miss your wheels-up time by a substantial margin they will have to call flow control in Washington D.C. to get you another time so it is relatively important to be ready when they are if you want to avoid further delay.

IFR Takeoffs

Launching an aircraft into instrument weather conditions without having your act together is never a good idea, and at a busy terminal area it's simply intolerable. It may be fine to get your departure routing messed-up a little bit leaving your uncrowded airport back home but the controllers at a big airport don't have the spare time to hold your hand while you sort things out.

A good mnemonic to use before you take off on instruments to make sure you have everything ready is **WART:** Weather, Abnormals, Route, & Terrain.

Weather is usually what you are thinking about before you take off into the soup anyway, but take a minute to *really* think about it. Is the weather you are about to operate in something you are truly prepared to handle?

We don't mean that just from the standpoint of ascertaining if it will be too rough up there for you to fly. You must think about other things also.

If the weather is so bad as you depart that it is unlikely that you could make a successful return to your departure airport you need to have an alternate in mind in case something goes wrong after takeoff. What if you lose an engine on a twin and have to return to the airport for an emergency landing? You will have your hands full enough already. If the weather is below landing minimums at your departure airport and you either don't have or can't get to a suitable alternate you are in deep trouble.

If you are flying a single engine aircraft IFR will you have enough time after breaking out of whatever weather you are flying into to make an attempt at a successful engine-out landing?

At busy airports the above questions get even stickier. If you are leaving a major airport and their weather is below landing minimums, it is unlikely any other airports in the immediate area will have more "landable" weather for two reasons: first, large airports usually have the best approaches and second, if the weather is really *that* bad it probably covers a wide area.

Part 121 of the FARs requires airlines to have a filed "takeoff alternate" when the weather is bad. It might not be a bad idea to at least have one in mind for yourself.

Abnormals covers three things, the aircraft you are flying, the airport you are flying it out of and yourself. You should clearly know the capability of the aircraft you are preparing to fly through the clouds of a major hub airport. Does all the equipment work? If not, what is left that you can use? Is there anything different or abnormal about the airport you are leaving today, such as out-of-service navaids?

Finally, you should consider any abnormalities in yourself. Are you feeling well today? If you have a severe head cold this is no time to be yanking and banking in those clouds.

Routing is vitally important to you if you are leaving a large, congested airport. More than likely you will be cleared out of the area using a SID (Standard Instrument Departure; more on these in the next chapter).

Of course, the most important thing about a SID is that you actually have the one you've been cleared to do with you. They are changed fairly often and a "Barnburner three departure" may be totally different from the "Barnburner two" that you have in your book. If you don't have the appropriate SID tell them and get alternate routing.

If you are cleared for a SID is it that important that you follow it? Only if you don't want to have a mid-air with an L-1011. Sometimes the routing out of a major airport is a very crowded thing and it is very important that you stay on course or at least tell someone if you can't.

Terrain. Go over major terrain features or problems on your instrument departure. The fact that you are going to be in the clouds and unable to see those hills and mountains makes them doubly dangerous to your longevity.

Also when you are thinking terrain, think density altitude. Some SIDs require a certain rate of climb and minimum altitudes for different segments...can your aircraft comply with them because of the field elevation and heat?

All of the above sounds complicated but becomes second nature after you work with it a while. Also, most of the time not all of the items have to be done. If you have flown out of an airport for some time you will have an idea of the terrain and SIDs you will have to be concerned with.

Turbulence On The Ground

The first place you will run into a wake turbulence problem today will be in the line for takeoff. If you get too close to that 727 you can get blown over when they put up the power to move up in line. If you stay too far behind, you will have everyone mad at you for wasting so much taxiway space.

Even a relatively light airliner like a 727 has to develop thousands of pounds of thrust to get moving from a dead stop on a taxiway.

Usually they will bring the power up quite a bit to get started then pull it back to almost idle once they get going. Your job is to survive the first power-up.

Knowing the correct safe distance is a matter of experience. Until you've been around long enough to know how far back is safe play it conservatively even if everybody else thinks you're a wimp.

Another problem with jet blast is that not only will it blow you over...it also stinks. Even airliners following airliners in a long line will sometimes turn their noses at an angle in order to spare their passengers jet fumes and spare the airliner behind them their own jet exhaust. Keep in mind that you and your passengers have better things to do than snort burning jet fuel.

Intersection Takeoffs?

Many times instead of making you wait behind a dozen jets, the ground controller will offer you an intersection takeoff.

This offer of an intersection wouldn't effect your spot in line usually. You'd still have to wait until the planes that were in front of you took off but at least you wouldn't have to smell the fumes and withstand their blast while you waited.

Is an intersection takeoff safe? That's a tough question. Most airline flight operations manuals forbid it following the old saying of "never leave runway behind you." Others allow an intersection takeoff in specific circumstances. Your decision should be based on a few other factors than legality.

At major airports the runways can be upwards of two miles long. Shortening the available space by 2,500 feet or so still leaves you with plenty of asphalt if something goes wrong. Still, every foot can count if you lose your engine at a bad moment.

Under most circumstances something like this would be perfectly safe and would probably be a lot safer than braving that wake behind the 727 for another fifteen minutes. The choice is up to you, the pilot in command. If you want full length nobody will be mad at you because you asked for it—but you may have to wait longer to depart.

Once it's your turn, the tower will clear you into position and tell you to wait for the wake turbulence of the airliner that just took off in front of you.

Many times in a busy airport ATC will only give you five miles behind a heavy to help things along. They will clear you for takeoff well ahead of the usual two minute wait assuming the heavy will be at least five miles ahead of you when you break ground. If you are flying a light airplane like a 310 do yourself a favor and ask for the full two minutes unless there is some other factor like a strong cross-wind that will help clear out the vortices for you.

In our example, Smith didn't need to worry about any of this. Though Burlington is an ARSA primary, there's not much traffic early in the morning. He was able to fire up and head straight out to the runup area.

Departure

Don't expect a turn on course until you are pretty well clear of the traffic pattern. Many times your departure routing doesn't make much sense in terms of your direction of flight. ATC is trying to work you through both their departure and arrival routing so many times their vectors seem illogical.

Usually at a Class B/TCA, they are in a big hurry to get rid of you if you are VFR...they have a pretty heavy workload handling the mandatory stuff, the IFR traffic. Because of this, you will probably only get a polite, "You are now leaving the TCA, squawk 1200 frequency change approved, good day."

If you want VFR flight following you will normally have to ask for it as they hand you off. ATC will bend over backwards to help you but if they are too busy with IFR traffic don't be surprised to have them tell you that they can't. Often if the controller isn't too busy he or she will suggest a new frequency without being asked. In any case, if you get a new frequency right away there's no need to squawk 1200—just keep the code and call the next controller. They'll assume you want to continue flight following until you cancel it, and you'll be handed off just as if you were on an IFR flight.

As always, the important thing about leaving congested airspace is to keep your eyes open. This is even more important to keep in mind because you will be spending much more time with your head inside the cockpit trying to make the government happy (maintaining heading, etc.).

Enroute

F or our purposes, a flight is considered to be "enroute" from the time it leaves the vicinity of the departure airport to the time it gets close to the destination. For all but uncontrolled airports, that means the edge of the surrounding airspace—Class B, C, or D.

For uncontrolled airports, you're enroute as soon as you've left the immediate vicinity of the airport.

Leaving the Area

When you're departing a controlled field IFR, departing the airspace is simple—just fly your cleared route and accept handoffs as they come. There's no real demarcation between the terminal area and the enroute environment. In many parts of the country, you'll be talking to approach controllers all the way to your destination.

Your workload is much higher at the start of your flight than it is after you're established enroute. You'll find that there's significantly more traffic close to the departure airport, particularly if it's a large one. Also, chances are you'll have a more complex navigation task close to the departure airport than farther out. On top of that, you need to pay attention to your airplane more until you're established at your cruising altitude. There are power settings to monitor, leaning of the engine, cowl flap settings, climb rates to monitor, and so on. As a result, you'll need to pay close attention to your duties until you get up to altitude and established enroute.

IFR Departure Routes

Major airports (those with heavy jet airline traffic) typically have separate departures for small and large airplanes to keep the fast traffic separated from the slow. Other airports might have only have a single departure.

A word is in order here on the difference between Standard Instrument Departures (SIDs) and IFR departure procedures. There's a lot of confusion among IFR pilots about these, and this confusion has resulted in accidents.

A SID is a "canned" departure route that gets an airplane from the runway to any one of several surrounding nav fixes, either navaids or intersections. It exists primarily to simplify clearance delivery, in the same way ATIS simplifies the transfer of current airport information. It also guarantees obstruction clearance. As often as not it's something as simple as "Fly runway heading. Maintain 3000 feet." Other SIDs are more complex, involving various altitudes, turns, and climb gradients. They're found on approach-plate-sized charts bound into NOS books or mixed in with Jeppesen plates.

Any IFR pilot can count on getting the SID when flying out of an airport that has one. Failure to specify it when filing the flight plan will often result in a full route clearance, even if the remainder of the route is acceptable to ATC. It is possible to specify "No SIDs" on your flight plan if your airplane is incapable of meeting its requirements, and ATC will give you a route that your equipment can handle. SIDs are also discretionary from ATC's point of view: they don't have to give you the route.

An IFR departure procedure is something different. It exists specifically to provide obstruction clearance for airports that have surrounding obstacles. FAA has established criteria for an imaginary plane that extends out and up, conelike, from an airport. If any obstacles poke up into this protected area, an instrument departure procedure is established, even if the airport doesn't warrant a SID. Unlike a SID, an IFR departure procedure is not discretionary. You must follow it if departing IFR, and it's a good idea to follow it even if VFR simply because it guarantees obstruction clearance.

IFR departure procedures are listed on Jepp plates on the same page as the IFR departure minimums—usually the bottom of the airport diagram page. In NOS books it's at the front of the book.

There's considerable confusion among both pilots and in some cases, FAA personnel about the difference between SIDs and IFR departure procedures. One well-publicized crash in 1991 involved a jet carrying country singer Reba McIntyre's band out of San Diego's Brown Field. The airport has an IFR departure procedure, but no SID. The departure procedure keeps pilots departing eastbound from running into the mountains.

The pilot was intending to depart at night, VFR, and pick up his IFR clearance enroute. He was cautioned about the departure procedure by the FSS briefer when he phoned in his flight plan: he said his charts were in the cockpit and he'd look it up when he got out to the airplane. Evidently he assumed the briefer was talking about a SID, because when he called back a few minutes later he told the briefer that he could find no mention of any special departure procedure in any of his charts. From the transcript of the conversation it's evident that the briefer wasn't too clear on this, either. Eventually the briefer wound up reading the procedure verbatim to the pilot. The pilot realized that if he were to fly that route he'd end up inside the nearby TCA. He didn't want to deal with getting a VFR clearance into the TCA, and so asked the briefer if he could just depart VFR eastbound at 3,000 feet. The briefer said that would be fine.

Apparently neither the pilot nor the FSS specialist understood the significance of the departure procedure. It contains a turn to the north to avoid the mountains to the east. Even though it's an IFR procedure and must be flown if operating IFR, merely flying VFR is not sufficient reason to ignore it. At night, the terrain may be all but invisible.

In any event, the pilot took off VFR, headed east, and within minutes had flown straight into the side of a mountain. Clearly this was the pilot's fault, since he should have had all the information necessary to tell that there were mountains there: however, it clearly shows the kind of confusion that exists about IFR departure procedures.

Together, SIDs and IFR departure procedures provide IFR pilots with guidance on how to navigate out of the terminal area safely and efficiently. As a side note, a VFR pilot can benefit from them (particularly SIDs) in that they show the route all departing IFR traffic will take when leaving an airport. If flying near an airport with a SID a VFR pilot can count on there being a lot of traffic in the vicinity of a charted SID route.

Leaving Controlled Fields IFR

The bigger the airport, the simpler (yet more hectic and demanding) departure is. As noted above, there's no real dividing line between the terminal environment and the enroute environment—the only real difference is the amount of traffic to worry about.

For airports in Class B/TCAs or Class C/ARSAs, you're considered to be enroute as soon as you leave the area, whether you've reached your cruising altitude or not. Typically it's signaled by a handoff to the next controller.

For generic controlled airports (Class D airspace) you're passed from the tower to a controller immediately after taking off. Here is where the enroute portion of the flight really starts, but as long as you're below 3,000 feet AGL or so you'll have to worry about traffic in the vicinity of the airport.

If you are flying a smaller and slower aircraft IFR out of a large airport be ready for some special handling. You will be using procedures slightly different from your larger, turbine powered relatives for a few very good reasons. The first has to do with speed and the second has to do with your altitude capability.

As noted above, many SIDs have performance requirements that are far beyond the capabilities of piston-powered airplanes. If you fly one of these departures in a small airplane ATC will turn you out of the normal jet traffic flow as soon as they can. Why?

In a TCA the maximum speed is 250 knots as opposed to the 200 knot limit in Airport Traffic Areas. Not only is this the maximum airspeed allowed, many times it is the airspeed that ATC wants to see to keep traffic separated and flowing smoothly. In TCAs, ATC expects you to accelerate to 250 knots pretty quickly and plans their departure traffic flow on the fact that you will. There's an obvious problem for them if your climb speed is 100 knots. Because of this you will probably be turned immediately after takeoff to a heading that will keep you out of the way of the jets behind you.

You want this, believe it or not. There is nothing to be gained by following a Boeing 757's path into the clouds if you don't have to. You will be risking a wake turbulence encounter if you do. If you take your own departure path, that is one less thing to worry about.

When you depart IFR from a major airport it is a very good idea to use the climb speed that will give you the highest airspeed and still give you an adequate rate of climb.

Weather avoidance is sometimes a problem for pilots not accustomed to operating in busy areas because they lack the self-confidence to demand deviations.

Departing in Bad Weather

If weather looks like it is going to be a problem along the route of flight that ATC has picked out for you don't hesitate to get on the radio and tell them you are deviating. It is common courtesy to request deviations from your cleared flight path and is also a regulation but sometimes it is better to inform rather than ask.

If you have to leave your last cleared route of flight you are required by FAR to get an amended clearance before you change course. There are a few problems with this approach. One is that if you ask for a deviation ATC may think you really don't need it and deny it. You might hear something like "One-three delta unable that left deviation for another fifteen miles." You don't want to hear that because you are looking straight at a level five thunderstorm.

It is almost impossible to be put in this kind of situation: usually ATC will grant your deviation, especially if you tell them there is no way you can continue safely on the course you are following.

If they don't allow the deviation you have no other recourse than to declare an emergency and go around the storm anyway. Under no circumstances should you allow yourself to be browbeaten into flying through a thunderstorm...the odds just aren't in your favor.

When you ask for a deviation and the controller comes back and denies you the heading you want but offers you another that doesn't have a traffic conflict, think twice before you reject it and go to the emergency position. The alternative may work out fine.

In crowded airspace you do have one luxury when it comes to thunderstorms and other severe weather. You are not usually the first person through it. There almost always is someone else that has flown through the area before you and you can use their experience as your guide. In the case of the alternate deviation suggested by the harried controller they usually will say something like: "One three delta, previous traffic reports only light chop and moderate rain through that area."

Even though another pilot's report is comforting and a good tool when you are trying to figure out which way to head in bad weather don't completely rely on them. The pilot that went through ahead of you may have a very high threshold of fear; moderate or worse turbulence may not bother him. Also, weather has a bad habit of changing very quickly. It may be much, much worse for you than it was for the person five minutes ahead of you.

Keep in mind that ATC sincerely wants to help you avoid the bad weather. In crowded terminal areas they don't have the time to help you that you might be able to get from Center. They have two things going against them. The first is that they have digital radar that is designed to blank out areas of weather so they can see the traffic more easily. The second is that their primary responsibility is to separate traffic, not be your personal weather advisor. In a pinch they will do their job—separating traffic. It is up to you to do your job and avoid the weather.

Leaving Uncontrolled Fields IFR

Non-towered airports are another matter when flying IFR. There are two options: if the weather is clear, you can choose to depart VFR and pick up your clearance at your convenience once airborne, or you can call ATC for a void time, take off and call ATC as soon as possible. In the former case you might have to hold while ATC finds room for you. In the latter you risk busting your void time.

In either case, you may have to be concerned with traffic in the vicinity of the departure airport. Many IFR pilots are in the habit of taking off and calling ATC immediately, while still quite close to the airport. If there's any traffic in the area, this can prove very dangerous, since it keeps you from learning about other aircraft arriving and keeps them from learning about you.

Unless you're absolutely sure there's no VFR traffic in the vicinity it's wise to stay on the unicom frequency and self-announce until well clear of the airport's vicinity—say, five miles and a couple of thousand feet above the traffic pattern. This way you can remain aware of anyone heading your way, inbound to the airport.

When you're ready to contact ATC, don't forget to make a call on unicom announcing that you're clear of the area. This lets everyone in the pattern know that you're gone, and they need not worry about you any longer.

Leaving Controlled Fields VFR

At Class B/TCA and Class C/ARSA airports, leaving VFR is much the same as leaving IFR. ATC cannot give you a SID, and you're not bound by IFR departure procedures, but you do have a clearance to follow until you leave the terminal area.

At generic towered airports, there's no clearance involved. After leaving the pattern all you have to do is wait until you're clear of the Class D/ATA before changing frequencies. Many pilots let the tower know they're clear of the area, but this is not required (it has the side effect of letting other pilots in the vicinity know that you've left the area, though).

Again, it's handy to have a copy of any applicable SID with you, so that you can avoid that route if operating in Class C/ARSA or Class D/ATA airspace. This takes you away from the IFR traffic, and also makes life easier for ATC: the controllers don't have to make room for you in their flow of IFR departures.

If operating VFR, a pilot has a choice to make when he or she reaches the edge of terminal airspace: whether or not to continue using ATC. We'll cover that a little later on.

Leaving Uncontrolled Fields VFR

The same caveats apply here as for IFR departures. A VFR pilot is perfectly within his or her rights to simply take off and switch frequencies the moment the wheels have left the ground, or for that matter without even turning on the radio. It may be legal, but it's not very smart.

The best policy it to treat a non-towered airport as if it were a towered one: don't change frequencies until you're clear of the area around it. When you do, tell everyone that you're clear.

At uncontrolled fields, you may find that it's necessary to climb to several thousand feet before you can even contact ATC or for them to be able to see you on radar.

Enroute IFR

The enroute portion of the flight is the quietest time for an IFR pilot. There's little to do other than navigate, monitor the weather and talk to ATC.

The actual process is simply an extension of getting out of the terminal area. Fly the cleared route, and accept handoffs from one controller to another. There are no complications with passing through terminal areas along the way or getting clearance into a terminal area at the destination: it's all taken care of by the IFR clearance.

Complications arise when the weather gets foul, and there are many fine books on how to deal with altering your route to deal with bad weather. That's not what this book is about, so we won't get into it here.

Enroute traffic for the IFR pilot in IFR conditions is taken care of by ATC. There's nothing the pilot could do about it in any case. In VFR conditions, however, things get more complicated.

Too many IFR pilots fly the same way on an IFR clearance when the weather is clear as when it's solid cloud. They spend all their time fiddling with charts and checking the gauges, when they should be looking for traffic as well.

Whenever there's the possibility of VFR traffic being in the area, an IFR pilot should fly as if he's VFR and spend most of his time looking out the window. Those VFR aircraft do not have to be in contact with ATC outside terminal airspace, and though they're supposed to fly at different altitudes than IFR traffic, they often do not.

Heavier traffic can be expected near enroute airports, and on heavily used routes. If you fly in a given area for a while, you can often get a feel for the more heavily traveled airways.

To avoid traffic, it's a good idea to go direct whenever possible. You'll need area navigation capability (RNAV or an IFR-certified loran or GPS receiver) if you want to go a long way direct and the permission of ATC, but it carries the dual benefit of less traffic and a more direct route.

ATC is often loath to give a direct route from the outset, but if you're flying a suitably equipped airplane you should not give up. Every time you get handed off to a new controller you should ask for a clearance direct. It's easy to do this without tying up the frequency. Here's a typical check-in with a new controller:

"Harrisburg approach, Cessna 1234 level at 6,000, looking for direct Westminster if possible."

This will be met with a warmer reception if you wait until you're far from the busy terminal area. If you were leaving White Plains, New York bound for Philadelphia, a request for a direct routing from the first controller you spoke to would result in a curt rebuttal: such a clearance would take you smack over the middle of Newark International. The controllers at major TRACONs have far to much to do to be accommodating special requests like that.

It's not even necessary to have area navigation equipment as long as you can navigate IFR to the fix you're requesting a clearance to. Many routes contain various kinks and doglegs that can be avoided if the controller grants your request of a direct routing.

ATC doesn't really care *how* you navigate from place to place, as long as you stay on course between the two points. You can dead-reckon if you really want to.

Our usual practice is to ask everybody we talk to for a direct routing as soon as we're clear of terminal airspace and the route wouldn't step on anyone's toes. Naturally, we also take into account terrain, large bodies of water and weather. More often than not we're able to save a considerable amount of time and distance.

Enroute VFR

The VFR pilot has much more freedom of action enroute when flying enroute than a pilot operating IFR. As long as airspace restrictions are obeyed, he or she can fly any route at any altitude without talking to ATC at all.

However, there are distinct advantages to using the system whenever possible, and flying logical, predictable routes.

The biggest safety advantage to using the system is traffic advisories. In crowded airspace it's very comforting to know that there's a controller out there who will call out traffic for you.

There's an interesting point to be made here: the regs say that ATC will provide traffic advisories to VFR traffic on a *workload permitting basis*. What, exactly, does that mean? We've all heard controllers turn down requests for traffic advisories because they were too busy, but what if they grant the request and issue a transponder code? Does that mean that the controller will call out every target for the VFR pilot, or only those he has time for?

The latter is the case. "Workload permitting" means that if it's convenient for the controller, the traffic will be pointed out to you— but there's no guarantee whatsoever. If another airplane zips by off your nose, don't complain to the controller about it: he's under no compulsion to call it out for you if you're VFR.

So, flight following is not a guaranteed safety net. Still, it's better than going without. We fly a great deal in the Northeast corridor, and on virtually every flight when we use flight following the controller calls out some traffic for us before we've seen it. Sometimes we never do see it.

The only traffic separation service guarantee there is applies to IFR pilots, who will have separation provided to them at all times. It's not legal for a VFR pilot to fly an IFR clearance in VFR conditions without an instructor, however (one more argument for getting your instrument rating).

Another advantage of staying close to the system (on airways, using VORs, etc.) is that it makes the job of search and rescue easier should you fail to show up at the other end of your trip. A clearly defined route can speed their task considerably.

Lastly, flying the airway system is more likely to take you close to airports along the way, which can be critical should you have an in-flight emergency.

Using our sample flight, Mr. Smith chose to use the system all the way from Burlington to LaGuardia, keeping to the airways and using flight following.

As he neared the edge of the Burlington ARSA, the controller told him that he was leaving the area, to squawk 1200 and that he could change the frequency. Rather than squawk 1200, then try to call another controller, Smith simply asked for flight following, and was granted it. No need to change squawks, and he was hooked into the system. The controller merely handed him off to the next controller, and so it went all the way to New York.

After reaching altitude and settling in, Smith told the controller he would be leaving the frequency for a couple of minutes to call Flight Service. The controller told him to report when back on.

The call to Flight Service provided him with the current and forecast LaGuardia weather, and allowed him to activate his flight plan. He provided the specialist with his actual departure time: had he not done so, the activation time would be the time he called, not the time he left, and his ETA would be fouled up.

The weather indicated that surface winds were northerly, and the forecast indicated that they were expected to remain that way. A check of the TCA chart and airport diagram suggested that LaGuardia's runway 4 would be in use.

As he passed out of Burlington's airspace, the controller suggested another frequency to try for flight following.

The only traffic to speak of occurred around the Albany ARSA. While Smith was overflying the ARSA, he was still within Albany's airspace and so talking to Albany approach. Four targets were called out, but he didn't see any of them.

In choosing to make use of the system, Smith increased his safety without significantly increasing his workload. His route took him over several medium-to-large airports and away from the mountains, and he had ATC helping him in the search for traffic. He also was able to keep up-to-date on the weather and even deduce a little bit about operations at the destination while still hundreds of miles away.

10

Arrival

Probably the most challenging part of flying into congested airspace is the arrival. If you're headed into a major airport you'll be trying to get to the same place as a lot of other airplanes, most of which a liable to be very large, very fast, and far less maneuverable than you are. If you're the typical GA pilot flying a single or light twin, you're definitely the odd man out here: big jets are the stock in trade of large airports, and your job is to fit into their flow as smoothly as possible.

If you get behind, even a little, you'll have a much harder time of it than if you keep on top of the situation at all times. That means having your charts ready, being sharp on the radio, and not allowing yourself to get overwhelmed by the fast pace and pressure.

Special Considerations

Major airports like O'Hare or Hartsfield cater to airline traffic. While you have a right to be there, their operations are not oriented towards you and you'll be required to fly your airplane differently than if you were flying into your local uncontrolled airstrip.

The biggest issue is speed. While there is a speed limit inside Class B/TCA airspace, it's far above the maximum cruising speed of most singles. Expect to hear a lot of admonitions to keep your speed up. You'll often find yourself flying down final as fast as your airplane can go. This is not difficult if you're prepared for it, and never forget that a few miles behind you is an airplane that stalls out at about your cruise speed.

This is not always the case. Many large airports have separate runways that are used for small airplanes, and this can take a lot of pressure off the GA pilot. But more often than not you'll find yourself sandwiched between "heavies" flying the ILS, and the last thing you want is to force one of them to break off the approach.

You also may be told to stay at an unusually high altitude in the pattern if VFR, or to fly unusually long legs.

Inside a Class B/TCA or Class C/ARSA communications are critical. It's of the utmost importance to listen for your call sign in all the chatter and answer promptly. You'll have a heavy workload at this time, so it's necessary to be even more aware of the radio than normal. It's also possible for the controller to drop the ball—we've been "lost" more than once while on vectors in IMC. If you keep up on your situational awareness, you can ensure that the controller is doing his part and that you're where you're supposed to be.

Another part of communication to watch out for in crowded areas is similar call signs. Once we were on vectors to an ILS in low IMC, expecting a turn in at any moment. The controller issued a vector that sounded right, but we weren't sure we'd heard our call sign. Just as we were about to call back to verify that the turn was for us, somebody else acknowledged the transmission. Since the controller didn't correct him, we concluded that the vector was for someone else. We kept on flying, right through the localizer, whereupon the controller called us and asked where we thought we were going. We told him we'd not been given a vector, and everything was straightened out...but we never did find out what happened to that other poor soul, who was obviously flying off in an unintended direction.

Lastly, remember that controllers are human, too, and can make mistakes. If you receive an instruction that clearly doesn't make sense, by all means have the controller clarify it. A typical example is being told to turn one way when it's obvious the controller should be telling you to turn the other.

Entering the Terminal Area IFR

When flying IFR your task is considerably different than when VFR. An IFR pilot doesn't need to worry about clearance into the Class B/TCA, since it's already taken care of by his IFR clearance. Instead, the IFR pilot needs to concentrate on setting up for, and executing, an instrument approach.

As with departure, there will be considerably more traffic the closer one gets to the airport. The type of traffic depends on the individual field: for example, a Class B/TCA primary will have very heavy airline traffic consisting almost entirely of large jets, while a reliever nestled under the edge of the same terminal area may have just a heavy a traffic load, only consisting of commuter and GA traffic. If you fly a piston-powered airplane, naturally it's a lot easier to merge with the flow at the reliever than it is at the primary.

As you begin to approach the edge of the terminal area, ATC is likely to give you a lower altitude. This is a good cue to tune your second comm radio into the ATIS for the destination, which you may be able to pick up from fairly far out. Listening to the ATIS on one radio while monitoring the other is tricky, but gets easier with practice. This is a place where a second pilot can help out, by listening to the ATIS on the cabin speaker and copying it down while you fly the airplane. Since you're outside the terminal area, the sector of airspace you're in is likely to be less crowded than those farther in, so there will be less radio traffic to interfere with picking up the ATIS. If you can't pick it up yet, try again periodically. If you find listening to a second radio to distracting, check that you can hear the ATIS, then ask the controller for permission to leave the frequency for a moment—this will be easier the farther out you are.

You'll be passed off from one approach controller to another, usually two or three times before you're with the final approach controller, the one who needs to know whether you've got the ATIS (there's no way of telling who's who other than through experience).

Somewhere along the way you'll be told which approach to expect. In VFR conditions, it will almost always be a visual. If it's not too busy, you can ask for the instrument approach in use to help satisfy your IFR currency requirements. If there's a second pilot, you can fly it under the hood, or if you encounter instrument conditions at any time during the approach, you can still log it as in instrument approach even if you were in the clouds for only a moment.

When you're told which approach to use, grab your approach plates and open them to the appropriate one. Scan it and note the important information: altitudes, headings, and frequencies. Set your radios and heading bugs appropriately.

Also at some point on the way in you'll probably be given a heading and altitude to fly. This marks the official end of the enroute segment of the flight, which means that you can stow your enroute chart and any other materials you've used during your trip. All you really need for the approach is the plate and a notepad. The earlier you get all this out of the way the better: things get more and more tight as you approach the airport, and the fewer distractions you have the easier the approach will be.

All during the arrival you should be watching for traffic if there's even a remote chance that there is any. While the controller is supposed to keep you separated from it, it's always possible that he doesn't see a target or something else has gone wrong. You have a lot to do at this point in the flight, but watching for traffic should still be very far up on the list.

You may well be asked to keep your speed up on the approach. In visual conditions this should not be a problem, but in IMC it's better to fly the approach by the numbers and let the following aircraft take care of themselves. Flying an ILS at an unusual speed is a good way to blow the approach.

Most of the mystery of an IFR approach is taken away when you fly into a big airport. There is always someone on the approach ahead of you and you can be fairly certain you'll see the runway and be able to land if dozens before you have.

Most major airports have some sort of parallel approach set-up. In Atlanta, for instance, all four of their runways are east-west and aircraft are making parallel approaches twenty-four hours a day.

Keeping traffic from flying into each other's final course is mostly a controller function. They even have a separate controller whose sole job is to monitor the final approach courses and make sure incursions don't happen.

Your job is to make absolutely sure that you know which runway you are shooting an approach for and that you stay on that final course, or if you wander off that it is to the side *away* from the parallel approach course.

Let's say you are cleared for the approach, roll in on final, shoot the ILS, get to minimums and see nothing. What kind of missed should you execute?

There are two kinds of missed approaches at big airports, the published missed and the missed you actually follow.

The missed approach for Atlanta Hartsfield's runway 9R tells you to "Climb to 1500', then climbing RIGHT turn to 3500' outbound via ATL VOR R-180 to SCARR INT/D15.0 ATL and hold."

This missed approach procedure works well and airline pilots do it every day...in the simulator. In real life, unless you just lost all communications with ATC it is very unlikely that you would shoot the missed as published.

In reality, after you missed and made your required radio call telling ATC of the fact you would probably be cleared straight out and vectored either for another approach or to a holding fix to wait the weather out or maybe even to a departure routing to send you on your way to an alternate airport.

Entering the Terminal Area VFR

Arriving at a large airport VFR is somewhat different than for IFR. The IFR pilot spends a lot of time preparing for the approach and executing it, while the VFR pilot has different tasks to perform.

Once again, let's look at our sample flight. About 15 miles north of Carmel VOR, his final reference point outside the New York TCA, Mr. Smith canceled flight following, called Flight Service and closed his flight plan. It no longer served any purpose, since he'd soon be under positive control for the rest of the flight, and it would be one less thing to worry about after landing.

He also stowed his enroute and sectional charts, and opened up the TCA chart, in keeping with the idea of making things as simple as possible inside the terminal area. Lastly, he tuned in the ATIS and copied it down. It noted that the ILS runway 4 approach was in use, so he opened his chart book to that plate and looked it over. Even though he was not instrument rated, he tuned in the localizer anyway: his airplane was equipped for it, and it would provide a handy reference for glideslope and runway alignment should he become disoriented.

By this time he was over Carmel, so he reduced power to keep from penetrating the TCA too soon and called approach. He was given a transponder code.

His clearance into the TCA included a heading and altitude, plus the runway to expect and the pattern leg on which he was to enter. He set the heading bug on his directional gyro, and used the ADF compass card as a reminder of the altitude.

From there on in it was a fairly simple matter of keeping a close watch on traffic, maintaining his heading and altitude, and complying with new instructions. This was much more hectic than he was used to.

Smith had to fly a high, fast downwind and a final approach at cruise speed to fit in between the DC-10s on final, but he was able to pull it off successfully.

Smith approached this flight with great thoroughness, and was well prepared for the experience even though he'd never flown into a TCA before. Many pilots, though, approach it "cold." On any given VFR day you can hear the shaky, stumbling voice of a nervous pilot trying to get through or around the TCA without the government coming down on him like a ton of bricks. The real difference is preparation and a good attitude in the cockpit (and by that we don't mean "upright and sitting in front of the yoke").

Arrival at Towered Airports

Certain towered airports that actually lie outside Class B/TCAs or Class C/ARSAs have special procedures that make arrival somewhat like penetrating the terminal area. For example, you may be required to contact approach before calling the tower. Often, this information isn't noted on the chart, but is listed in the Airport/ Facility Directory. If you call the tower by mistake, you'll be given the appropriate frequency and told to stay away.

For the vast majority of towered airports, however, arrival is simple. For IFR pilots, it's nothing more than being handed off to the tower controller after establishing yourself on the approach. For VFR pilots it's contacting the tower before entering the Class D/ATA.

As with terminal areas, it's a good idea to stow all your enroute materials, close your flight plan if VFR, and get the ATIS before calling the tower. The more crowded the airport, the more important preparation is.

You should be aware that many towered airports make use of local landmarks that may not be marked on charts, so if the controller tells you to use one you're not familiar with, tell him so immediately.

For IFR pilots, in VFR conditions you'll probably be issued a visual approach. Since the average towered airport is likely to have a very full pattern on a clear day it's a good idea to just accept it—practice approaches and full traffic patterns don't mix very well.

If you're flying VFR you can select your altitude. It's a good idea to descend to near pattern altitude before entering the Class D/ATA. This will make it easier to spot traffic in the pattern, since it won't be lost in ground clutter, and will make you easier to see as well. It's important, particularly at crowded airports, to fly your pattern at the proper altitude and heading. Often the pattern will get so full that four or five mile downwind legs are necessary. A pilot making a straight-in who's a little off to one side and a pilot flying downwind who's also a bit off course can come uncomfortably close to a head-on collision (it happens).

Arrival at Non-Towered Airports

Uncontrolled fields can be downright dangerous on clear days, with a lot of casual "Sunday pilots" out for some landing practice, instrument students conducting practice approaches and commuter flights all mixing it up together.

The same guidelines used at towered airports apply here as well. Fly the correct headings and altitudes, and be predictable. Keep your patterns square.

Communications are just as important here as in a Class B/TCA, only in this case it's self-announcing that's important.

Making use of standard procedures should keep you out of trouble, but a few words are in order if you're practicing IFR approaches into an uncontrolled field on a VFR day.

Self-announcing is just as important here as anywhere, and you should call out your position at every critical point of the approach along with your intentions if appropriate. Remember that many of the pilots you're warning aren't instrument rated, and may not know where the outer marker lies or even what it is. Some examples:

"Oxford traffic, Cessna 1234 is seven south of the field on a practice ILS 36 approach, and we're showing a landing light."

"Oxford traffic, Cessna 1234 is eight south, turning inbound on the ILS 36 practice approach."

"Oxford traffic, Cessna 1234 is on the ILS 36 practice approach three miles out, and we have the Piper on base in sight."

"Oxford traffic, Cessna 1234 is a mile south, breaking off the approach to the right, we'll be conducting another one."

When you break off the approach it should always be outside the traffic pattern, and if possible you should turn away from it no matter what the missed approach procedure says. Be sure your safety pilot checks for others making non-standard base leg entries: if there's someone there, you'll be turning right into them.

Trip Summary

Of all the portions of a flight, the arrival segment is the most demanding. The pilot has a great deal to do, and is under pressure to make no mistakes at all.

Proper preparation before reaching this stage of a flight into congested airspace is the real key to making it to a safe arrival.

11

Safety Techniques

Even though flying is increasingly gadget-oriented, with moving map displays, sophisticated navigational instruments and even collision-avoidance devices for general aviation pilots, the fact remains that collision avoidance itself is still predicated on the concept of "see and be seen."

You may be flying the latest in technology in a Boeing 767, using the flight management system, the autopilot and auto throttles. When traffic arises on the controller's scope they will still call it out to you in the form of a visual reference to an analog clock face: "Traffic two o'clock, a mile and a half, altitude unknown."

Collision-avoidance devices like TCAS and Ryan's TCAD notwithstanding, you're still bound to rely on your eyes when it comes to staying away from other airplanes.

Eyes

There is nothing that can or will ever compare to your eyes when it comes to detecting, judging and avoiding other aircraft in crowded skies. In your repertoire of survival tools in congested airspace your eyes are the most important.

Like all tools, your eyes are prone to error and misuse. If you have a weather radar on board and don't know how to properly use it or aren't aware of some of its built-in errors you are probably better off not using it at all. The same thing applies to your eyes.

One of the most important traits of your eyes is that in order for them to see something properly you must be looking directly at the object in question. You certainly may see that aircraft approaching you from the side using your peripheral vision, but gauging its distance and motion accurately are not really possible unless you look directly at it.

Images moving across your field of vision are very hard to see correctly and almost impossible to judge. For that reason a quick scan for traffic using fast head movements is pretty much a waste of your time. You will either miss the traffic completely or misjudge it.

It is true that your eye can scan a 200 degree field in a glance. The problem that arises is that your eye only has a very small part of the retina, in the center, called the fovea that has the ability to send sharp, focused images to your brain. All the other information that is not processed by the fovea will be less detailed and clear to your brain in its interpretation of the data. For example an aircraft observed by the fovea that is seven miles away would have to be only 7/10 of a mile away for the brain to recognize it if it is out of the foveal range.

According to the FAA the best way to scan outside your aircraft is to take it in short, regularly spaced eye movements that bring in areas of the sky, one after the other into your central visual field. They recommend that each eye movement should be limited to about ten degrees and each area should be observed for at least one second to give you a better chance to see something.

Most experts recommend that you scan outside your aircraft using a "sectors" method. Look at one section of the sky at a time letting your eyes focus on just that sector. Try to work out your own habit pattern, one that will allow you to make a complete scan of the sky on a regular basis but still allows enough time for your eyes to get the whole picture.

How Much Time to Spend Looking?

Just about every pilot and book on flying agrees that everybody should spend a lot of time looking out of the windows of their aircraft searching for traffic. From their very first flying lesson pilots are taught to "keep their neck on a swivel" and point out to the instructor any other airplanes they may see even if they think the instructor already sees it.

However, flying an aircraft in crowded airspace such as a Class B/TCA or Class C/ARSA requires a certain amount of precision. You'll be required to maintain certain altitudes and headings even if flying VFR. This will require you to keep your eyes in the cockpit more than you usually would under more relaxed conditions—at the very time when you are close to a lot of traffic.

This is becoming especially true of the newer types of airliners present in the crowded airspace you will be flying in. Although they do have sophisticated autopilots and automatic flight systems of all types the point remains that they must be programmed. The programming keypads are, naturally, located inside the cockpit in a place where one or more of the pilots must look down and concentrate on putting in the correct data.

Although heads-up displays (HUDs) are currently in use in certain military aircraft and are on the near horizon for airliners and general aviation aircraft, for the time being every time you want to see what your aircraft is doing you must put your attention and eyesight back inside the cockpit.

FAA studies have shown that the time most pilots spend looking at things inside their cockpits should be about 1/3 to 1/4 of the time spent looking outside the aircraft, with scans of the instruments limited to four or five seconds.

This may not represent the real world. We've been guilty of spending long periods of time staring at the instruments especially in nice, VFR weather and suspect most others do the same. While it will probably drive you crazy to figure out what percentage of your time is spent peering at the outside world, it is a good idea to work on spending as much time as you can looking for other aircraft.

Good Scanning

Most people are comfortable with looking at things from left to right so it is probably a good idea for you to get in the habit of starting your scan on your left shoulder and working your way to your right. Take your field of vision a little bit at a time and take a little time to let your eyes focus on what you are looking at.

If you take a quick look outside at one ten degree area and then look quickly back in at the instruments your eyes will probably tire quicker than they would if you do a larger outside scan first.

The important thing about all this talk of scanning is to do what works best for you. You can get a bad headache just thinking about scanning techniques. Just be aware that especially in crowded airspace, your eyes are your life.

Night Vision

When it's dark outside you face certain problems with your eyesight and must deal with them in order to enhance your chances of survival.

Obviously unless the other aircraft are sporting very visible lighting such as strobes, landing lights or the "billboard" lights on the vertical stabilizers of airliners you will have a more difficult time seeing them at night. Another problem is that it is harder for your eyes to judge their distance and direction at night.

In a way, though, seeing traffic while flying at night is much easier than it is in daylight. It's much easier to pick out a red flashing beacon against a dark sky than it is to see the tiny speck of a Cessna 172 three miles away in daylight. This is balanced against the near-instrument conditions found on dark nights. In many areas of the country, you might as well be inside a cloud as flying VFR on a dark night: the horizon can be virtually invisible.

Aircraft lights can cause all sorts of illusions. It is not uncommon at all at night to think you are on a collision course with another aircraft only to find out it is six miles away. This is particularly true when an airplane showing a landing light is headed right at you. It's amazing how close a light can seem on a clear night, even though it may be ten miles off.

Small print and colors on charts and aircraft instruments are more difficult to see during dim conditions. Without adequate cockpit lighting seeing things in the aircraft is impossible, with it you run into another vision problem: dark adaptation.

Dark adaptation is the reason that many aircraft still have red instrument lights. In order to let the eye adapt to dim lighting outside it takes up to thirty minutes of complete darkness. Most pilots can achieve some degree of "night vision" after twenty minutes exposed to only very dim red instrument lights.

This was valuable for flying at night in World War II but has very little value in terms of your survival in the night skies of the 1990s. As a matter of fact using only red instrument lighting has the drawbacks that you can't properly read charts and instruments and also may lead to disorientation—something you definitely don't want while flying.

What to do? Like most other things in flying and life you have to achieve some sort of balance. Keep the instrument lights dim enough to maximize your outside vision but bright enough to read the instruments and fly the aircraft.

Empty Field Myopia

When your eyes sense that there is nothing to see they automatically go into a sort of automatic "relaxation mode."

Let's say you are flying VFR above an overcast or a large body of water. With nothing specific for your eyes to work on they tell themselves it is time to take a break. They will automatically relax and adjust themselves to a comfortable focal distance of ten to thirty feet. This is extremely bad news. Since you're no longer focused at infinity traffic appearing in your field of vision may not be noticed until it is uncomfortably close to you.

The way to combat this "auto laziness" is to keep an active scan going. Try to keep your eyes moving around the cockpit and outside the aircraft. If you get in the habit of just staring at the horizon over the glare shield you may be guilty of looking without seeing.

VFR Collision Avoidance

Assume you've just spotted another aircraft. How can you tell if you are on a collision course with it? There are two components to this, relative altitude and heading.

If the traffic in question appears to be below the horizon it is probably below you. If it is above the horizon it is probably above you and if it is right on the horizon it is probably at the same altitude as you. Simple, but there are a few problems here.

A single glance at the approaching traffic may give you the false impression that you are safe. The aircraft may be climbing or descending into your altitude. Unfortunately there really is no way to figure out if the other airplane is climbing or descending unless you watch it for a while and notice its movement relative to the horizon. In the real world, though, you may have more on your mind than that one aircraft. Other traffic may be called out to you by ATC, you will be busy flying the airplane and if you are in a Class B/TCA, you will be expected to fly it with a relative degree of accuracy. All of that means more "heads down" time.

Another problem with aircraft close to the horizon has to do with seeing them in the first place. An airplane from above or below is a fairly sizable object, while one seen edge-on (particularly from ahead or behind) is essentially just a line. That line can blend very easily with the horizon and simply disappear.

The second component of a collision course is relative heading. The old maxim says, "If it stays in one spot on your windshield you are on a collision course."

This is true, most of the time (it could be headed away from you). Rest assured that if you see another aircraft and it is staying in the same spot on your windshield, and it is getting larger, you are in very serious trouble and should get out of there.

It may be difficult for you to see the most threatening aircraft, that one that isn't moving on your windscreen or field of vision. Your eye will tend to ignore a stationary target...it prefers to track a moving one. Unless the target is directly in the center of your line of sight your eye may ignore it completely until it is very, very close.

Maintain Contact

It's all too easy to lose a target once you've spotted it, especially if it's slightly below you silhouetted against the ground or if in-flight visibility isn't what it could be. If you see some traffic, then lose it you're right back where you started.

It is, of course, impractical to keep looking at the other airplane. The best way to maintain contact with the other airplane while still managing your other tasks is to note its position relative to the background, so you'll know roughly where to look the next time you search for it.

Evasive Action

The FAA, in its recommendations, suggests that you should remember your right-of-way rules and try to follow applicable FARs when trying to avoid that other aircraft.

That's fine as far as it goes, but when you get an unexpected windshield full of 727, the last thing you should do is try to figure out which side you should pass on.

Think about it while you're reading this. You are in the air. You have just been surprised to discover an Aztec appear in the lower left of your windshield. It is large and is getting larger; your mind will only have about a second to register this. If you are flying some kind of light twin your closure rate is probably somewhere around three hundred and sixty knots. If you noticed this traffic when it was a quarter mile from you that means you have only approximately 2.5 seconds until you are dead. Figure that it takes your airplane roughly a second to do what you tell it to do...especially if you are on autopilot. That leaves you with a second and a half to decide what to do to avoid the collision.

This one and a half seconds will probably not be long enough for your brain to look up the pertinent FARs and right of way rules. You will be depending mostly on reflex action in this case to get you away from the threat.

There are two things to burn into your memory, however:

• **Change altitude.** An airplane can change altitude much faster than it can change heading. Also, you only need to change altitude by a few feet to miss an airplane at the same altitude as you are. Your very first reaction if you find yourself eyeball-to-eyeball with someone you didn't see is to change altitude away from them. (They're unlikely to be at *exactly* the same altitude, and you're likely to do the right thing instinctively.)

• **Turn towards the other airplane.** Unless you're head-on or close to it, turning towards the other airplane will cause you to miss it, because it will no longer be there once you arrive. If you turn away, you may wind up crossing his flight path with disastrous results. This, of course, depends on circumstances—just remember that he's moving, too, just as fast or faster than you are.

Overstressing Your Aircraft

While certainly possible, it's unlikely that you'll yank on the yoke hard enough to damage your airplane, particularly if you're at or below maneuvering speed. Most potential accidents like this don't require that much of a correction to avoid hitting the other airplane.

Obviously, yanking your aircraft into some sort of aerobatic maneuver that you can't recover from is probably not a huge improvement over a mid-air but most likely you won't allow yourself to do that. And if it is necessary to do some scary stuff to avoid that big bang you are much better off being alive trying to recover than dead and spiraling in.

Multiple Threats

In most cases, there's a little more time to actually think things out. This is when you should remember your right of way rules. There is nothing like having adequate time to help you make the right decisions. One of the first things you should consider when looking at traffic you may have trouble with, especially if you do have time to think about it, is the fact that there are probably other airplanes in the vicinity.

It won't do you much good to avoid one hunk of traffic only to blunder into another that you weren't thinking about. This is especially true in congested environments like traffic patterns. A little awareness of where you are and what you are trying to do will go a long way to help you survive.

For example, if you are flying the visual approach to 8R in Atlanta and you see some traffic pull in front of you that you want to avoid you would know that it would be much safer to break right than to break left. Why? Because there are aircraft shooting approaches parallel to yours on runway 9L just to your left. It won't do you much good to avoid one conflict only to throw yourself into ten others.

Clearing Turns

Almost anybody that has ever taken even one flying lesson is familiar with the concept of clearing turns. If you had a good instructor he or she stressed the importance of avoiding collisions from the very first day of learning.

Before starting just about every maneuver, you were taught to make a series of two clearing turns of at least a ninety degree heading change, one to the left and then one to the right. This was done to make sure that before you started that stall series, that you would avoid any inadvertent matings with other aircraft.

Making two ninety degree turns won't do in a TCA environment, but you can still look under or over that wing before you begin any heading changes. It doesn't take much. On a high-wing airplane, just lift the wing on the side you're getting ready to turn up a few degrees and peek under it. This is also a good time to glance in the direction away from the turn. If you are VFR in a very congested piece of airspace chances are that everybody is flying faster than you. That airplane that you're turning away from may be closing in on you at over 200 knots.

Be sure, once again, while making these turns to look both above and below your aircraft. Remember, airplanes do not have to be level with you to be a threat.

IFR Collision Avoidance

While being "in the soup" technically relieves you of worrying about having to dodge VFR traffic, this attitude can lead you to some serious trouble.

The first faulty assumption many pilots make is that everybody follows the rules. The fact is that there are some people out there who fly their aircraft through clouds and TCAs without clearances illegally every day.

When trying to avoid a collision with these people, once again the best tools in your survival tool chest are your eyes.

This seems a paradox. How can you see them if they're in the clouds with you? Well, consider: Just how far *can* you see when you are actually flying through clouds?

Think about it for a minute. You are in the air, cruising in a solid cloud bank at six thousand feet. Is it really solid? You can see your wing tip without distortion. How much further can you see in this cloud? You're surrounded by whiteness but are you actually still *in* the cloud? Could you just be in between two indistinct layers?

The point is, you don't really know because you have no real frame of reference. You might think you can only see two feet in front of your windshield but may actually have 3/4 of a mile. You have no way of knowing unless there is some physical object out there that you can judge by, such as another aircraft.

If there ever was a case for empty field myopia, this is it here in the clouds. Not only are you conditioned as a pilot not to peer out of the windows when you think you are in the clouds, your eyes also conspire to rob you of some of your observation talent when you may need it most.

There have been many occasions when we were in IMC, thinking we were enclosed by clouds and still saw other aircraft as they cruised by at least a thousand feet away vertically. As a matter of fact, this happens all the time in situations like IFR holding patterns. The point is, you just don't know how much you can see unless you look out of the window.

It isn't easy to do this outside observing because almost all of your training as an instrument pilot has negated most of the "keep your head on a swivel" training you got as a pre-solo student.

When you flew with a hood on to get your training remember how every time you raised your head your instructor would slam it with a rolled-up chart and yell "no cheating!"? Well, it had its place when the intent was to keep you from figuring out if you were blue side up by looking outside but in that swat was another, more tacit lesson. It was: *real* instrument pilots don't look outside of their aircraft for a minute.

The second and more sobering lesson you learned as you started out that keeps your eyes inside your aircraft today was that while in the clouds if you looked outside too long when you came back in you were disoriented. Remember that? Remember staring outside too long when you were in actual conditions and when you glanced at the instruments you were in a thirty degree bank that you didn't realize that you were in?

These two lessons that have been reinforced by years of trying to keep your altitude and heading and course right on the money have probably led you to the habit of not looking outside your aircraft very much when you are operating IFR—as often as not, even in VFR conditions.

This habit can be very dangerous. Not only are there still people out there who will fly through cloudy weather without a clearance or any kind of contact with ATC you can also be slammed by perfectly legal pilots operating in a legal manner.

Most of the time when you are operating your aircraft on an IFR clearance in congested airspace the weather is at least basic VFR. Although you did get the instrument rating in order to fly through just about any kind of weather, for the majority of the time you are flying VFR on an IFR clearance.

Unless you are flying above flight level 180 and are operating in Class A/Positive Control Airspace you will have to face the fact that you are sharing the skies with VFR pilots that may or may not be in any kind of contact with ATC.

Also, unless you are flying above 12,500 feet MSL or near a terminal area, even if the controller points out traffic to you as you fly in and out of the clouds it is unlikely you will get any altitude information. Remember, altitude encoding transponders are not universally required.

Sometimes the "mix" can get pretty hairy. Even if everyone is following the FARs to the letter you can have some close calls.

Let's say you are flying along at 12,000 on an IFR clearance. You are flying on the centerline of the airway, are exactly on altitude and have been listening to the melodic voice of the controller as he talks to other traffic. You are flying in and out of a broken overcast that starts at about your altitude.

Suddenly you see another aircraft headed right for you. It is a DC-3, it is *huge* and it is headed right at your windshield. What should you do? Of course, you yank the aircraft up to the left to avoid the collision. Your passengers get the thrill of experiencing three or more gees and you get ATC's attention as you scream about your near-miss to them.

Of course, you weren't really going to collide with that DC-3. It was also legal, perhaps a shade above the correct VFR altitude of 11,500 feet but still uncomfortably close to you. The pilot of the Douglas in question was following the applicable FARs to the letter. The weather was nice so he decided that filing IFR would be a hassle. He decided to fly airways back to the home base and chose to cruise at a VFR cruising altitude of 11,500, perfectly legal for his direction of flight. (He was too close to the cloud bases, but remember our earlier discussion of how difficult it is to judge distance from clouds? He may have believed he had the required separation.)

Although he probably should have been talking with ATC just as a precaution he decided, quite properly, not to. All the controller could see on his scope that might have helped you avoid this scare was a VFR transponder code return of 1200. Normally he would have called this out to you but his workload was heavy at the moment so he didn't...nor was he required to.

Popping out of that cloud an eighth of a mile from the oncoming DC-3 you reacted just as you were designed to do. You demonstrated a classic "flight response" and you got out of there, fast.

Although ATC is required to point out other traffic to you, the IFR pilot, when you are operating in such airspace as TCAs and ARSAs their definition of pertinent traffic will usually vary with their workload.

If there is no apparent conflict they may fail to tell you anything at all about traffic that you may consider to be a threat.

The only way to protect yourself when flying IFR during VFR conditions is to keep your eyes open and outside the cockpit as much as you do when you are flying VFR. Remember, ATC is only charged with separating you from other IFR traffic when you are flying on an IFR clearance. They will try to point out VFR traffic to you but you are really on your own.

Before we move on, you should be aware that the situation just described is a rare one. Many VFR pilots die because of incursions into IFR conditions (it's the number-one cause of fatal accidents for many types of aircraft), but it's almost never because of conflict with IFR aircraft. Much more dangerous to the IFR pilot are common, everyday invisible killers like icing conditions, embedded thunderstorms and severe turbulence.

Taxi Through The Fog

Collisions aren't just a problem in the air. Poor visibility on the ground has caused a few spectacular and tragic air carrier accidents as well. Two of the best examples are the mid-'70s collision at Tenerife in the Canary Islands in which hundreds died and the 1990 runway incursion at Detroit that resulted in the collision of a DC-9 and a 727.

Probably the hardest controlling job in existence is that of ground controller at a busy, congested airport. Traffic is much closer together, the potential for collision is greater than that in the air they are all operating at the same altitude (ground level) and there is little or no radar available to monitor traffic on the ground. About the only advantage a ground controller has over someone working traffic in the air is that he or she can actually ask the traffic to come to a complete stop.

You, as a pilot during IFR conditions face just as many problems and dangers as the ground controller; even more if you consider the fact that it is you that are the one that is going to be in that potential collision.

If you think it would be difficult to taxi around Chicago's O'Hare, imagine doing it with one half mile visibility in snow and fog—at night.

Becoming disoriented while taxiing can be very hazardous to your continued flying health. That's what happened in the 1990 Detroit accident: the DC-9 pilot didn't know he was on the runway.

One of the main problems during low visibility conditions is that the control tower cab is usually well into and sometimes above the fog layer and the controller has absolutely no view of what he or she is controlling. They can only control the airport by referring to little "position reports" that you are able to give. In other words, if you tell the controller you are on alpha taxiway and are actually just about to forcibly meet a heavy DC-8 on the runway you just blundered onto, you are on your own...the controller actually thinks you *are* on taxiway alpha.

It would be nice while you are taxiing around in the fog to have a good mental picture of the airport you are moving around on. If you operate at that airport on a regular basis, this is no problem and you can probably taxi around it with your eyes closed. If you're not a "local" take a few minutes before you start the engine to look over the taxi chart for that airport. Of course, you can't memorize an airport in just a moment of study, but you can get a general idea of how things are laid out.

Another taxi hazard will occur with both IFR and VFR pilots. If you taxi directly into the afternoon sun you are blind and can stumble into all kinds of trouble if you aren't looking out and thinking ahead.

When in doubt about where you are while taxiing or where you are going or even how you are supposed to get there use the one advantage you have over pilots that are in flight. Stop your aircraft right there (unless you are about to be smashed by a 747) and tell the controller you need help.

While we're on the subject of the dangers of taxiing, when a controller clears you to a runway does that mean you can cross all the other runways to get there? According to the FAA, anytime a ground controller tells you to taxi to a runway, you are cleared to cross all runways that your taxi route intersects except the assigned runway. If they want you to hold short of anything else they are supposed to tell you.

Because of the inherent confusion, ATC never utters the word "Cleared" when they give you taxi instructions. That is to preclude you thinking that you have been "cleared" into position on the active runway. Their instructions to taxi always say something like: "Cessna one-two bravo, taxi to runway one-four. Hold short of runway two-seven at taxiway alpha, acknowledge hold."

Although some airports request that you read back all runway holding instructions, most do not. It is still a very good idea to get in the habit of reading the holding instructions back to the controller. It is professional, backs you up and is a very good habit. Again, if there is any doubt about crossing any runway or taxiway, stop and ask.

One of the last things you need if you are flying a smaller aircraft IFR out of a busy airport is to be turned over by jet blast. It happens to the unwary. Give jets—even small ones—a wide berth while you are taxiing. This may be difficult if you are waiting in line for takeoff but try to give yourself as much room behind them as you can. It may be a good idea to accept that intersection takeoff just to avoid the stuff. Another thing to remember is that anytime an airliner or air transport has its rotating beacon on it is running engines, so look out.

Congested Areas

Since most airplanes still navigate in reference to VORs you may find a lot of traffic crossing directly over them. Keep firmly in your mind that even though you are flying VFR, many IFR flights may be cleared over the same fix you are over. For example, Rome, Georgia is one of the primary fixes on the arrival at Atlanta's Hartsfield International Airport. Almost every bit of traffic flowing into Hartsfield from the northeast flies over this VOR. You are perfectly within your rights to fly VFR over this fix also at any legal VFR cruise altitude you choose. The VOR lies outside the Atlanta TCA and you don't even have to be in radio contact with ATC to fly there, nor do you need a transponder (Rome is beyond the 30-mile Mode C "veil").

It is a fact of life, though, that turbojets will be flying over Rome VOR, five miles in trail at about 13,000 feet MSL by the dozens at any given time of day. The minimum enroute altitude for the Rome Arrival is 4,000, so it wouldn't be surprising at all to see dozens of lighter aircraft coming over this fix too.

You can see that even if you weren't going to have anything to do that day with Atlanta, such as if you were planning a flight from Chattanooga to Montgomery, you still might find a world of trouble in terms of traffic when you got near Rome VOR.

Even if you are planning to land at Hartsfield, are in radar contact with ATC and are doing everything they say you are still in greater danger over a heavily traveled fix like Rome. Why? Because, as we saw earlier, VFR pilots may operate over this VOR any time they like, at any altitude they like and may still not be on radar.

There are other places besides VORs that you should use caution. A good example of this would be the Miami, Florida area just west of the cities of Fort Lauderdale, West Palm Beach and Boca Raton. If you fly in this area you are probably familiar with the "canal," a large north-south drainage ditch just west of the city and Interstate 75 and the Florida Turnpike.

Just to the west of this line lies the Everglades, to the east lie some of the busiest airports in the southern United States. Most local VFR pilots in the area know that if they stay west of that line they are going to be pretty much free of the heavy traffic going into Miami International and Fort Lauderdale. The traffic for these airports usually comes from the west because the prevailing wind is usually from the east, but they almost always cross this canal at 3,000 feet or above. If you were to wander a few miles east of this line and tried to fly VFR north or south you would be up to your eyeballs in 727s, DC-9s and L-1011s.

Likewise, just about any coastline or beach can be counted on to have more than its fair share of traffic. Pilots, particularly those of piston singles, don't like to fly over open water, and for good reason. They'll hug the coastline, often just for the view. It's amazing how close to the beach most pilots will stay—even a mile offshore or inland can result in considerably less traffic.

Make It Easier On Yourself

Some of the most effective techniques are the simplest:

• **Have a clean windshield.** Is that an F16 coming right at you or is it that bug you hit on final two weeks ago? Simple, and you can save yourself many frightening experiences if you just clean your windshield from time to time. If you can't see out properly because the windshield is dirty, or covered with last weeks bug collection think how stupid you'll feel when you do kiss nose to nose with a *real* F16 that you thought was a dead moth!

• **Don't block your own vision.** There are enough natural obstructions to your vision like door posts, struts, wings and engines without adding to them yourself. There is no problem if you want to stack all of your charts on the glareshield for a minute but don't leave them there all day. Same thing goes for hats, coffee cups, or any of the other common cockpit flotsam. If it is between your eyes and the

windshield or window, it is in your way and is a potential safety hazard.

• **Use those lights.** Even in broad daylight, your landing lights can save your life. Just about any exterior light at your disposal can be used to make you more visible to other pilots any time of the day. They may be just the thing that catches the other pilot's eye that's on a collision course with you. Definitely use all the lights that you have available. This includes strobes, your rotating beacon, and especially landing lights.

Most airline pilots and other pilots of large aircraft use all their landing lights any time they are below ten thousand feet, no matter the time of day. It makes them ten times more visible to the other pilots. There is no reason why this technique shouldn't work on all aircraft, especially those that are smaller or may not have strobes. Every little thing that can make you more visible to the other pilot may save your life.

• **Utilize labor-saving devices.** Your autopilot and other labor-saving devices could save your life because they free your mind for the all-important chore of avoiding becoming tomorrow's headline. It has already been mentioned that right when you should be looking outside your aircraft the most you are required by ATC and the FARs to fly the airplane more precisely in terms of heading and altitude. If you have an autopilot that will hold altitude or even just heading for you while you peer outside you will have that much more time to search for traffic that you would have spent staring at the altimeter or heading indicator. When you enter congested airspace, leave your ego at home if you want to survive. If there is a tool or piece of equipment on board that will help you do the job easier and safer, use it.

• **Use your passengers as tools.** Unless you fly an airliner or a big corporate jet you are probably in close contact with whatever passengers you are carrying into this area crowded with other airplanes. They all have eyes too, so why not use them as an added survival tool?

There is no need to scare everybody on board with horror stories. Just simply, as you make your usual pre-takeoff brief about seat belts and emergency exits, include this phrase: "If any of you see any other aircraft around us while we're flying be sure to quietly point them out to me, even if you think I already see them."

Your passengers will all probably be looking out the windows anyway. Might as well use those extra eyes to enhance your safety. You must ask them, though. You'd be surprised how many passengers think either that you already see the traffic they see or will get mad at them if they mention it to you.

One resourceful pilot we know devised a clever game to keep kids occupied in flight. He put a 25¢ bounty on any airplane one of the kids could spot before he did. Cheap insurance, and it gives them something to do.

Out-Fake The Wake

If it hasn't happened to you yet, it is bound to sooner or later if you fly around large, crowded airports very long. You will find yourself out on final some fine day and the controller will say, "Piper two-seven alpha, cleared to land runway two-six right, wind calm, caution wake turbulence, Boeing 767 five miles ahead landing."

You just might feel a large bump and then your world will try to turn upside-down. That's wake turbulence.

This isn't just a problem for light airplane drivers. We've been in both DC-9s and 727s that have gone through some very interesting rolling maneuvers behind heavies.

If you are flying a light airplane remember that for you, everything bigger than you is a "heavy." The FAA only recognizes aircraft that have a certificated gross weight of over 300,000 pounds as a "heavy" and although they will usually warn you about the others they aren't under the same onus to provide spacing on them for you.

A fully grossed-out Boeing 727 can weigh over 182,500 pounds. A DC-9 can weigh over 100,000. Just about any airliner, jet or transport can cause you serious problems.

A little-thought-of, but very lethal form of wake turbulence is put out by helicopters. You don't expect such a small aircraft to cause much of a problem, but try landing or taxiing by one that is near the ground in a hover.

A few things to remember about wake turbulence:

• **It can kill you.** If you get in the wing tip vortex of a large aircraft you may experience a roll rate of over 720 degrees per second. That's better than twice the maximum roll rate of most production aircraft.

• **It can hang around.** Although the vortices do sink, they do so slowly in calm air. Also, if there is a crosswind they can blow over from a parallel runway and knock you flat.

• **Fly high.** If you can fly above the flight path of the heavy you should avoid the turbulence. This is easy on landing, especially if you have an ILS receiver and the runway has that approach. If there is a glideslope available to a runway the airlines are required by FAR to be on or above it. They'll usually be on it, so your safety factor should be to stay above it a little. This will put you above the wake. You will land a little bit long if you do this but you are in a light aircraft and the runway was long enough to land the heavy, it is probably very long.

On takeoff you will have a problem because it is very unlikely you will be able to out-climb something like a Boeing 767. In this case ATC will make you wait either two minutes or until the heavy is at least five miles away from you when you leave the ground. Remember they will not give you that separation from something like a 727 or DC-9 unless you ask for it.

Asking for more time behind a heavy is no crime. The controllers would rather see you cause a small delay than the long one that would happen if you crashed because of a wake turbulence encounter. Since, at the larger airports, they'll probably have you going off an intersection it is no problem for them in terms of traffic flow. If you think you need more time, just ask for it. They can't make you take off if you don't want to.

Along with the usual cautions about landing and taking off around heavies you should also be careful about crossing their wakes. You can get quite a jolt from them although they're not life threatening.

If you do find yourself in the clutches of wake turbulence it is important to treat it like an emergency, because it is one! Get out of there. If you are on short final it is unlikely you will complete the landing anyway so go around. In any case, get out of it the best way you can. If you are on final you can probably climb out of it. It is doubtful you can do that on takeoff. The best thing is to turn out of it as soon as you can, keeping in mind that there may be a lot of other traffic present. Don't press your luck with wake turbulence. It can kill you.

If you're IFR, you have a special set of problems. Even though you are in the clouds and can't see the heavy aircraft five miles ahead of you the problem of avoiding its wake turbulence still exists.

Have you ever noticed that when there is a discussion about how to recognize and avoid wake turbulence it's always assumed you can see the aircraft generating the wake you're trying to avoid? There are always pictures, for example, telling you to be sure to notice where the "heavy" touches down on the runway so you can land a little past that point to avoid the wake.

The trouble is, when you are shooting an ILS approach to minimums there is absolutely no way to see where the thing hit the ground.

It looks good on paper to say that you should avoid being behind and below a heavy aircraft but how do you plan to do that if both of you are in the clouds?

Wake turbulence is something you very much want to avoid if you plan to survive in crowded airspace. Even if you are the world's best recoverer from unusual attitudes you probably will have little chance of walking away from a full-blown wake encounter if you are in the clouds when it happens. Add to the extreme roll rate generated by a vortex the fact that most general aviation artificial horizons will tumble when they reach eighty to ninety degrees of bank and you can see the obvious problem. Are you interested in recovering from inverted flight in the clouds on a needle, ball and airspeed?

All you can really do is avoid it in the first place. Make it very clear to ATC that you want maximum spacing behind a heavy. Remember that just about any transport or airliner is a heavy to you if you are flying a light aircraft and that ATC will only give you special heavy separation from aircraft who can weigh 300,000 lbs or more. A 757 can cause a Cessna 182 driver very serious problems even though a 757 isn't considered a heavy.

Enroute or in the approach or departure pattern make sure that ATC knows you don't mind a small delay if it will give you more distance from a heavy aircraft.

On takeoff take the full delay that ATC offers you for heavy separation. Also, if they'll give it to you, request an initial heading away from the path of the big guy. This may be part of the departure anyway.

On approach fly a dot high on the ILS and land a little bit long. Most runways at large airports are long enough for you to get away with landing a few hundred feet long and it shouldn't cause any problems.

ATC Assistance

We've already covered VFR flight following elsewhere, and we recommend you use it whenever you can. Of course, the busiest times are when you're most likely to need it, and when it will be least available. In that case you're on your own.

ATC can provide many other kinds of assistance beyond flight following, mostly related to emergencies. The most common is offering vectors to the nearest airport.

If the spinach hits the fan and you need to land *right now,* by all means tell someone! The accident record is full of pilots who failed to yell for help until it was too late. If you're not in contact with ATC, squawk 7700 and give a Mayday or Pan-Pan call (Pan-Pan means an urgent situation that hasn't escalated to a full-blown emergency) on 121.5. Be absolutely sure to give your location. If you crash, they'll know where to look, and if you don't, they won't waste time trying to find out where you are in order to lend assistance. Also tell them your problem, and don't worry about etiquette.

This is another reason to use flight following. You're already in contact with ATC, and they already know where you are, so there's no need to switch to 121.5 and start over. Simply tell the controller you've got a problem. They'll probably put you on another frequency for handling.

If you are in radar and radio contact, don't waste time with charts. Just follow the controller's instructions and head for that field while running through your emergency checklists.

If your engine does quit, a good rule of thumb is to look down about 40-45 degrees from the horizon. You'll probably be able to safely glide to any field in that circle, so pick one and don't be tempted to push your luck.

ATC can also offer help for lost pilots. This has saved people on occasion, but there are also instances of pilots abusing it in the same way as people call 911 to complain about the noisy party next door. When you need help, call, but if you really don't, help yourself.

Lost Communications IFR

Every IFR pilot can quote chapter and verse about what the FAA expects them to do if they lose all their communications radios. Almost every IFR pilot recites these comforting regs while forgetting two very important things:

1. If they lose all communications capability it is very likely that they have lost all or most of their navigation capability as well.

2. If you lose comm in VFR conditions or subsequently find VFR conditions along your route you are expected to **land as soon as practical**. "As soon as practical" pretty much leaves the ball in your court.

If you are flying a jet, landing on a three thousand foot long grass strip is not a good idea. Passing up a five thousand foot paved airport and re-entering the clouds isn't a good idea either. The regulations leave it totally up to you.

If you are enroute, on your way to a major airport and you lose your communications radios let's look at your choices.

While you are within your legal rights to enter a TCA even if you do encounter VFR conditions (remember, the regs say you must land if it is *practical*) it may not be a good idea either for you or for ATC.

According to ATC whenever you lose communications with them, in addition to assuming you are going to follow the comm-out rules, they also make it a point to keep all the other traffic out of your way. This is because of the little clause that says you still retain your emergency authority to do whatever you think is best for the safety of your flight...including breaking any FARs you feel it is necessary to secure that safety. (You'd better be prepared to justify yourself when the FAA comes calling after you land.)

Let's say you have no communications. Let's also say that your destination weather at New York's LaGuardia airport is two hundred overcast with one mile visibility in a thunderstorm. You are still one hundred miles or so away from LGA and you are flying over a thin overcast that bottoms out around four thousand feet.

Most people would say that the most legal thing to do would be to fly to LaGuardia using the route you were last cleared on, shoot the approach to the runway you think is the active at or after your scheduled arrival time, land and go the coffee shop for a snack.

Legal? Yes. Good idea? No.

If you have also lost your navigational capabilities it is obvious that it is a real emergency and that following any kind of "cleared" IFR route is out of the question. What should you do then?

Head for where you think you can find VFR conditions and keep your eyes open for help. When most people fly IFR they don't think about where the weather is good. They usually are concerned with where the weather is bad. *Always* keep in the back of your mind which direction you would have to fly to get an improvement in the weather.

If the weather you are currently flying into is based on a cold front, for example, then your best route of escape should you lose communications and navigation would be to the northwest.

Keep your eyes open for other help that may be on the way. ATC will be tracking you and it won't take long for them to figure out that you are having problems. Use other resources that you may have to get help. Try the voice feature of any radios you may have left like your ADF or VOR. Pull out your handheld (you do have one, don't you?). Do you have a mobile telephone on board? Use it!

Have an Attitude

As a last word on safety techniques, we stress that you should develop a good attitude. There have been far too many accidents that could have been avoided had the pilot simply taken positive, assertive action in time to save himself. Getting "behind the power curve" is a sure route to disaster.

There have been even more accidents that could have been avoided had the pilot merely taken a moment to think before doing something unwise. Too many accident reports leave you asking yourself, "What was this guy *thinking* about?"

The most pertinent part of a good attitude when it comes to crowded airspace is the idea of defensive flying. Always assume that the controller doesn't see all the traffic on his screen, that the other guy doesn't see you, and that whoever you're looking at is not going to follow the rules.

We've tried to include in this book some sound strategies that will get you through crowded airspace, but they'll only work if you make them work. Fly within your capabilities and make use of all the tools available to you, and you should have no trouble at all.

Glossary

Note: Most of these definitions are based on the FAA's compilations in the FARs and in the Airman's Information Manual. Many of them have small additions or revisions the author has made in order to make them simpler or easier to understand. Also, certain terms that don't appear in any official glossary but, nevertheless, are used every day by pilots and controllers...slang if you will, but still very much in use. Pilots are encouraged to refer to current FARs and the AIM for up-to-date definitions and cross-reference information.

Abbreviated IFR flight plan—This is pretty much the official term for "air-filing" an IFR flight plan and obtaining a clearance. Using this method, ATC may authorize an IFR clearance by requiring the pilot to submit only the information necessary for the purpose of air traffic control. Usually the information required is only the type of aircraft, its location and the pilot's request. The controller may request other information for separation or control. This method is also commonly used for obtaining a climb to VFR on top. It should be noted that this is not the best way to get a clearance and ATC can only grant it when their workload permits. They may, if too busy to handle you, tell you to file through flight service.

Abeam—In relation to a navigational fix or physical point your aircraft is "abeam" when it is 90 degrees to your track. The term "abeam" is a very general one; you can be abeam a fix and still be 100 miles away from it.

Acknowledge—This is a term almost never used in flight. It means "did you receive and understand my message?"

Administrator—The Federal Aviation Administrator or any person to whom he has delegated his authority in the matter concerned.

Advisory frequency—The radio frequency used for airport advisory service such as Unicom, Enroute flight advisory and Flight Watch.

Affirmative—Yes.

Aircraft approach category—A grouping of aircraft based on a speed of 1.3 times the stall speed in the landing configuration at maximum gross landing weight. An aircraft shall fit in only one category. If it is necessary to maneuver at speeds in excess of the upper limit of a speed range for a category, the minimums for the next higher category should be used. Here is a run-down of the categories:

Category A—Speed less than 90 kts.
Category B—Speed 91 kts or more but less than 121 kts.
Category C—Speed 121 kts or more but less than 141 kts.
Category D—Speed 141 kts or more but less than 166 kts.
Category E—Speed 166 kts or more.

Aircraft classes—For the purposes of wake turbulence separation minima, ATC classifies aircraft as heavy, large and small as follows:

Heavy—Aircraft capable of takeoff weights of 300,000 pounds or more whether or not they are operating at this weight during a particular phase of flight. In other words, the aircraft must only be able to be that heavy. It may not actually weigh 300,000.
Large—Aircraft of more than 12,500 pounds maximum certificated takeoff weight, up to 300,000 pounds. Quite a range of weight. A large aircraft can be anything from a DC-3 to a Boeing 757.
Small—Aircraft of 12,500 pounds or less, maximum certificated takeoff weight.

Air Defense Identification Zone (ADIZ)—The area of airspace over land or water, extending upward from the surface, within which the ready identification, the location and the control of aircraft are required in the interest of national security.

Airman's Information Manual (AIM)—A publication containing basic flight information and ATC procedures designed primarily as a pilot's instructional manual for use in the national airspace system of the United States.

AIRMET—(Airman's Meteorological Information) In-flight weather advisories issued only to amend the area forecast concerning weather which is of operational interest to all aircraft and potentially hazardous to aircraft having limited capability because of lack of equipment, instrumentation, or pilot qualifications. AIRMETs concern weather of less severity than that covered by SIGMETs or Convective SIGMETs. AIRMETs cover moderate icing, moderate turbulence, sustained winds of thirty knots or more at the surface, widespread areas of ceilings less than 1,000 feet and/or visibility less than 3 miles and extensive mountain obscurement.

Airport advisory area—The area within ten miles of an airport without a control tower or where the tower is not in operation and on which a Flight Service Station is located.

Airport advisory service/AAS—A service provided by flight service stations or the military at airports not serviced by an operating control tower. This service consists of providing information to arriving and departing aircraft concerning wind direction and speed, favored runway, altimeter setting, pertinent known traffic, pertinent known field conditions, airport taxi routes and traffic patterns, and authorized instrument approach procedures. This information is advisory in nature and does not constitute an ATC clearance.

Airport/Facility Directory—A publication designed primarily as a pilot's operational manual containing all airports, seaplane bases, and heliports open to the public including communications data, navigational facilities, and certain special notices and procedures. This publication is issued in seven volumes according to geographical area.

Airport surface detection equipment/ASDE—Radar equipment specifically designed to detect all principal features on the surface of an airport, including aircraft and vehicular traffic, and to present the entire image on a radar indicator console in the control tower. Used to augment visual observation by tower personnel of aircraft and/or vehicular movements on runways and taxiways.

Airport surveillance radar/ASR—Approach control radar used to detect and display an aircraft's position in the terminal area. ASR provides range and azimuth information but does not provide elevation data. Coverage of the ASR can extend up to 60 miles.

Airport traffic area/ATA—Unless otherwise specifically designated in FAR part 93 that airspace within a horizontal radius of 5 statute miles from the geographical center of any airport at which a control tower is operating, extending up to but not including, an altitude of 3,000 feet above the elevation of an airport. Unless otherwise authorized or required by ATC, no person may operate an aircraft within an airport traffic area except for the purpose of landing at or taking off from an airport within that area. ATC authorizations may be given as individual approval of specific operations or may be contained in written agreements between airport users and the tower concerned. **Note: the dimension is five *statute* miles, not nautical. It is one of the few things in aviation still predicated on statute mileage.** This is replaced under ICAO airspace by Class D, which will typically be five nautical miles in radius.

Air route surveillance radar/ARSR—Air route traffic control center (ARTCC) radar used primarily to detect and display an aircraft's position.

Air traffic clearance/ATC clearance—An authorization by air traffic control, for the purpose of preventing collision between known aircraft, for an aircraft to proceed under specified traffic conditions within controlled airspace.

Air Route Traffic Control Center/ARTCC—An Air Traffic Operations Service facility consisting of four operational units.

> 1. Central Flow Control Function/CFCF—Responsible for coordination and approval of all major intercenter flow control restrictions on a system basis in order to obtain maximum utilization of the airspace. (see Quota Flow Control)
>
> 2. Central Altitude Reservation Function/CARF—Responsible for coordinating, planning and approving special user requirements under the Altitude Reservation (ALTRV) concept. (see Altitude Reservation)
>
> 3. Airport Reservation Office/ARO—Responsible for approving IFR flights at designated high density traffic airports (JFK, LaGuardia, O'Hare, and Washington National) during specified hours.
>
> 4. ATC Contingency Command Post—A facility which enables the FAA to manage the ATC system when significant portions of the system's capabilities have been lost or threatened.

Alert notice/ALNOT—A request originated by a flight service station (FSS) or an air route traffic control center (ARTCC) for extensive communication search for overdue or missing aircraft.

Altitude readout/automatic altitude report—An aircraft's altitude, transmitted via the Mode C transponder feature, that is visually displayed in 100 foot increments on a radar scope having readout capability.

Altitude reservation/ALTRV—Airspace utilization under prescribed conditions normally employed for the mass movement of aircraft or other special user requirements which cannot otherwise be accomplished. ALTRVs are approved by the appropriate FAA facility.

Altitude restriction—Altitudes specified by ATC to be flown until reaching a certain point or time. They are issued usually due to terrain, traffic or other airspace considerations. They are usually called crossing restrictions by pilots.

Approach gate—An imaginary point used within ATC as a basis for vectoring aircraft to the final approach course. the gate will be established along the final approach course one mile from the outer marker on the side away from the airport for precision approaches and one mile from the final approach fix on the side away from the airport for nonprecision approaches. In either case, the gate will be no closer than five miles from the landing threshold.

Area navigation/RNAV—A method of navigation that permits aircraft operation on any desired course within the coverage of station-referenced navigation signals or within the limits of a self-contained system capability. The major types of equipment are:

> VORTAC referenced or Course Line Computer (CLC) systems, which account for the greatest number of RNAV units in use. To work properly, the CLC must be within the service range of a VORTAC.
>
> OMEGA/VLF although two separate systems, can be considered as one operationally. A long-range navigation system based upon Very Low Frequency radio signals transmitted from a total of 17 stations worldwide.
>
> Inertial (INS) systems, which are totally self-contained and require no information from external references. They provide aircraft position and navigation information in response to signals resulting from inertial effects on components within the system.
>
> MLS Area Navigation (MLS/RNAV), Area navigation using MLS ground facilities as a reference.
>
> LORAN-C is a long-range radio navigation system that uses ground waves transmitted at low frequency to pro-

vide user position information at ranges of up to 600 to 1,200 nautical miles at both en route and approach altitudes.

Global positioning system (GPS) receivers, which use the network of GPS satellites for three-dimensional position references of startling accuracy. The great advantage of GPS is cost: a sophisticated unit with a full database costs less than $2,000 installed and can navigate better than far more expensive and complicated systems of only a few years ago.

ATC clears—Used to prefix an ATC clearance when it is relayed to an aircraft by other than an air traffic controller.

Automatic altitude reporting—That function of a transponder which responds to Mode C interrogations by transmitting the aircraft's altitude in 100 foot increments.

Automatic terminal information service/ATIS—A continuous broadcast of recorded noncontrol information at certain locations. This is done to relieve both the controller and the pilot of having to make repetitive transmissions of routine information.

Cardinal altitudes or flight levels—Odd or even thousand foot altitudes or flight levels. For example 6,000, 9,000, FL 270.

Ceiling—The heights above the earth's surface (AGL) of the lowest level of clouds that is reported as "broken," "overcast" or "obscured" that isn't considered partial or thin.

Center weather advisory—An unscheduled weather advisory issued by Center Weather Service Unit meteorologists for ATC use to alert pilots of existing or anticipated adverse weather conditions within the next 2 hours. A CWA may modify or redefine a SIGMET.

Charted VFR flyways—These are flight paths that are recommended for use to bypass areas heavily traveled by large turbine-powered aircraft. This is a voluntary program. VFR Flyway Planning charts are published on the back of existing VFR Terminal Area Charts.

Charted visual flight procedure (CVFP) approach—And approach wherein a radar-controlled aircraft on an IFR flight plan, operating in VFR conditions and having an ATC authorization, may proceed to the airport of intended landing via visual landmarks and altitudes depicted on a charted visual flight procedure.

Class A airspace—The ICAO designation for what had been the Positive Control Area.

Class B airspace—The ICAO designation for what had been a Terminal Control Area (TCA).

Class C airspace—The ICAO designation for what had been an Airport Radar Service Area (ARSA).

Class D airspace—The ICAO designation for what had been an Airport Traffic Area (ATA). Also replaces the Control Zone (CZ).

Class E airspace—The ICAO designation for what had been controlled airspace.

Class G airspace—The ICAO designation for what had been uncontrolled airspace.

Clearance limit—The fix, point, or location to which an aircraft is cleared when issued an air traffic clearance.

Clearance void time—Used by ATC to advise an aircraft that the departure clearance is automatically canceled if takeoff is not made prior to a specified time.

Cleared as filed—Means that the aircraft is cleared to proceed in accordance with the route of flight filed in the flight plan. This clearance does not include the altitude, SID, or SID transition.

Cleared for (name of) approach—ATC authorization for and aircraft to execute a specific instrument approach procedure for that airport: e.g., "Cleared for ILS runway 36 approach."

Cleared for approach—ATC authorization for an aircraft to execute any standard or special instrument approach procedure for that airport. Normally, an aircraft will be cleared for a specific instrument approach procedure.

Cleared for takeoff—ATC authorization for an aircraft to depart. It is predicated on known traffic and known physical airport conditions.

Cleared for the option—ATC authorization for an aircraft to make a touch-and-go, low approach, missed approach, stop and go, or full stop landing at the pilot's discretion. It is normally used in training so that the instructor can evaluate a student's performance under changing conditions.

Cleared to land—ATC authorization for an aircraft to land. It is predicated on known traffic and known airport conditions.

Climb to VFR—ATC authorization for an aircraft to climb to VFR conditions within a control zone when the only weather limitation is restricted visibility. The aircraft must remain clear of clouds during the climb.

Closed traffic—Successive operations involving takeoffs and landings or low approaches where the aircraft does not exit the traffic pattern.

Clutter—In radar operations, clutter refers to the reception and visual display of radar returns caused by precipitation, terrain, numerous aircraft targets, or other phenomena. Such returns may limit of preclude ATC from providing services based on radar.

Codes/transponder codes—The number assigned to a particular multiple pulse reply signal transmitted by a transponder.

Conflict alert—A function of certain air traffic control automated systems designed to alert radar controllers to existing or pending situations recognized by the program parameters that require the controller's immediate attention and/or action.

Conflict resolution—The resolution of potential conflict between aircraft that are radar identified and in communication with ATC by ensuring that their radar returns do not touch. Pertinent traffic advisories shall be issued when this procedure is in effect.

Contact approach—An approach wherein an aircraft on an IFR flight plan, having an air traffic control authorization, operating clear of clouds with at least one mile flight visibility and a reasonable expectation of continuing to the destination airport in those conditions, may deviate from the instrument approach procedure and proceed to the destination airport by visual reference to the surface. This approach will only be authorized when requested by the pilot and the reported ground visibility at the destination airport is at least one statute mile.

Controlled airspace—Airspace designated as a control zone, airport radar service area, terminal control area, transition area, control area, continental control area and positive control area within which some or all aircraft may be subject to air traffic control. **Note: This definition will change when the ICAO airspace designations go into effect. See Chapter 3.**

Controlled departure time (CDT) programs—These programs are the flow control process whereby aircraft are held on the ground at the departure airport when delays are projected to occur in either the enroute system or the terminal of intended landing. The purpose of these programs is to reduce congestion in the air traffic system or to limit the duration of airborne holding in the arrival center or terminal area. A CDT is a specific departure slot shown on the flight plan as an expected departure clearance time (EDCT).

Control sector—An airspace area of defined horizontal and vertical dimensions for which a controller or group of controllers has air traffic control responsibility, normally within an air route traffic control center or an approach control facility. Sectors are established based on predominant traffic flows, altitude strata, and controller workload. Pilot communications during operations within a sector are normally maintained on discrete frequencies assigned to the sector.

Control slash—A radar beacon slash representing the actual position of the associated aircraft. Normally, the control slash is the one closest to the interrogating radar beacon site. When ARTCC radar is operating in narrow band (digitized) mode, the control slash is converted to a target symbol.

Convective sigmet/WST/convective significant meteorological information—A weather advisory concerning convective weather significant to the safety of all aircraft. Convective SIGMETs are issued for tornadoes, lines of thunderstorms, embedded thunderstorms of any intensity level, areas of thunderstorms greater than or equal to VIP level 4 with an area coverage of 4/10 (40%) or more, and hail 3/4 inch or greater.

Coordinates—The intersection of lines of reference, usually expressed in degrees/minutes/seconds of latitude and longitude, used to determine position or location.

Coordination fix—The fix in relation to which facilities will handoff, transfer control of an aircraft, or coordinate flight progress data. For terminal facilities, it may also serve as a clearance for arriving aircraft.

Cross (fix) at (altitude)—Used by ATC when a specific altitude restriction at a specified fix is required.

Cross (fix) at or above (altitude)—Used by ATC when an altitude restriction at a specified fix is required. It does not prohibit the aircraft from crossing the fix at a higher altitude than specified; however, the higher altitude may be one that will violate a succeeding altitude restriction or altitude assignment.

Cross (fix) at or below (altitude)—Used by ATC when a minimum crossing altitude at a specific fix is required. It does not prohibit the aircraft from crossing the fix at a lower altitude; however, it must be at or above the minimum IFR altitude.

Cruise—Used in an ATC clearance to authorize a pilot to conduct flight at any altitude from the minimum IFR altitude up to and including the altitude specified in the clearance. The pilot may level off at any intermediate altitude within this block of airspace. Climb/descent within the block is to be made at the discretion of the pilot. However, once the pilot starts descent and verbally reports leaving an altitude in the block, he may not return to that altitude without additional ATC clearance.

Further, it is approval for the pilot to proceed to and make an approach at the destination and can be used in conjunction with:

> • An airport clearance limit at locations with a standard/ special instrument approach procedure. The FARs require that if an instrument letdown to an airport is necessary, the pilot shall make the letdown in accordance with a standard/special instrument approach procedure for that airport, or
> • An airport clearance limit at locations that are within/ below/outside controlled airspace and without a standard/special instrument approach procedure. Such a clearance is not authorization for the pilot to descend under IFR conditions below the applicable minimum IFR altitude nor does it imply that ATC is exercising control over aircraft in uncontrolled airspace; however, it provides a means for the aircraft to proceed to the destination airport, descend, and land in accordance with applicable FARs governing VFR flight operations. Also, this provides search and rescue protection until such time as the IFR flight plan is closed.

Decoder—The device used to decipher signals received from ATCRBS transponders to effect their display as select codes.

Delay indefinite (reason if known) expect further clearance (time)—Used by ATC when there is a delay of some kind but it is impossible to give the pilot an accurate estimate of the delay. For example, a disabled aircraft on the runway or weather below minimums.

Direct—Straight line flight between two navigational aids, fixes, points, or any combination thereof. When used by pilots in describing off-airway routes, points defining direct route segments become compulsory reporting points unless the aircraft is under radar contact.

Discrete code/discrete beacon code—As used in the Air Traffic Control Radar Beacon System (ATCRBS), any one of the 4,096 selectable Mode 3/A aircraft transponder codes except those ending in zero zero; e.g., discrete codes: 0010, 1201, 2317, 7777; non-discrete codes: 0100, 1200, 7700. Non-discrete codes are normally reserved for radar facilities that are not equipped with discrete decoding capability and for other purposes such as emergencies (7700), VFR aircraft (1200), etc.

Discrete frequency—A separate radio frequency for use in direct pilot-controller communications in air traffic control which reduces frequency congestion by controlling the number of aircraft operating on a particular frequency at one time. Discrete frequencies are normally designated for each control sector in enroute/ terminal ATC facilities. Discrete frequencies are listed in the Airport/Facility Directory.

Distance measuring equipment/DME—Equipment (airborne and ground) used to measure, in nautical miles, slant range distance of an aircraft from the DME navigational aid.

DME fix—A geographical position determined by reference to a navigational aid which provides distance and azimuth information. It is defined by a specific distance in nautical miles and radial, azimuth, or course (i.e., localizer) in degrees magnetic from that aid.

DME separation—Spacing of aircraft in terms of distances determined by reference to distance measuring equipment (DME).

DoD FLIP—Department of Defense Flight Information Publications used for flight planning, en route, and terminal operations. FLIP is produced by the Defense Mapping Agency for world-wide use. United Stabs Government Flight Information Publications (en route charts and instrument approach procedure charts) are incorporated in DoD FLIP for use in the National Airspace System (NAS).

Emergency locator transmitter/ELT—A radio transmitter attached to the aircraft structure which operates from its own power source on 121.5 mHz and 243.0 mHz. It aids in locating downed aircraft by radiating a downward sweeping audio tone, two to four times per second. It is designed to function without human action after an accident.

En route air traffic control services—Air traffic control service provided to aircraft on IFR flight plans, generally by centers, when these aircraft are operating between departure and destination terminal areas. When equipment, capabilities, and controller workload permit, certain advisory/assistance services may be provided to VFR aircraft.

En route automated radar tracking system/EARTS—An automated radar and radar beacon tracking system. Its functional capabilities and design are essentially the same as the terminal ARTS IIIA system except for the EARTS capability of employing both short-range (ASR) and long-range (ARSR) radars, use of full digital radar displays, and fail-safe design.

En route flight advisory service/Flight Watch—A service specifically designed to provide, upon pilot request, timely weather information pertinent to his type of flight, intended route of flight, and altitude. The flight service stations providing this service are listed in the Airport/Facility Directory.

En route minimum safe altitude warning/EMSAW—A function of the NAS Stage A en route computer that aids the controller by alerting him when a tracked aircraft is below or predicted by the computer to go below a predetermined minimum IFR altitude (MIA).

Execute missed approach—Instructions issued to a pilot making an instrument approach which means continue inbound to the missed approach point and execute the missed approach procedure as described on the Instrument Approach Procedure Chart or as previously assigned by ATC. The pilot may climb immediately to the altitude specified in the missed approach procedure upon making a missed approach. No turns should be initiated prior to reaching the missed approach point. When conducting an ASR or PAR approach, execute the assigned missed approach procedure immediately upon receiving instructions to "execute missed approach."

Expect (altitude) at (time) or (fix)—Used under certain conditions to provide a pilot with an altitude to be used in the event of two-way communications failure. It also provides altitude information to assist the pilot in planning.

Expected departure clearance time/EDCT—The runway release time assigned to an aircraft in a controlled departure time program and shown on the flight progress strip as an EDCT.

Expect further clearance (time)/EFC—The time a pilot can expect to receive clearance beyond a clearance limit.

Expect further clearance via (airways, routes or fixes)—Used to inform a pilot of the routing he can expect if any part of the route beyond a short range clearance limit differs from that filed.

Expedite—Used by ATC when prompt compliance is required to avoid the development of an imminent situation.

Fast file—A system whereby a pilot files a flight plan via telephone that is tape recorded and then transcribed for transmission to the appropriate air traffic facility. Locations having a fast file capability are listed in the Airport/Facility Directory.

Feeder fix—The fix depicted on Instrument Approach Procedure Charts which establishes the starting point of the feeder route.

Feeder route—A route depicted on instrument approach

procedure charts to designate routes for aircraft to proceed from the enroute structure to the initial approach fix (IAF).

Filed—Normally used in conjunction with flight plans, meaning a flight plan has been submitted to ATC.

Filed en route delay—Any of the following pre-planned delays at points/areas along the route of flight which require special flight plan filing and handling techniques:

• Terminal Area Delay—A delay within a terminal area for touch-and-go, low approach, or other terminal area activity.

• Special Use Airspace Delay—A delay within a Military Operations Area, Restricted Area, Warning Area, or ATC Assigned Airspace.

• Aerial Refueling Delay—A delay within an Aerial Refueling Track or Anchor.

Final—Commonly used to mean that an aircraft is on the final approach course or is aligned with a landing area.

Final approach course—A published MLS course, a straight line extension of a localizer, a final approach radial/bearing, or a runway centerline all without regard to distance.

Final approach fix/FAF—The fix from which the final approach (IFR) to an airport is executed and which identifies the beginning of the final approach segment.

Final approach-IFR—The flight path of an aircraft which is inbound to an airport on a final instrument approach course, beginning at the final approach fix or point and extending to the airport or the point where a circle-to-land maneuver or a missed approach is executed.

Final approach point/FAP—The point, applicable only to a non-precision approach with no depicted FAF (such as an on-airport VOR), where the aircraft is established inbound on the final approach course from the procedure turn and where the final approach descent may be commenced. The FAP serves as the FAF and identifies the beginning of the final approach segment.

Flight level—A level of constant atmospheric pressure related to a reference datum of 29.92 inches of mercury. Each is stated in three digits that represent hundreds of feet. For example, flight level 250 represents a barometric altimeter indication of 25,000 feet; flight level 255, an indication of 25,500 feet.

Flight plan—Specified information relating to the intended flight of an aircraft that is filed orally or in writing with an FSS or an ATC facility, or via computer and modem with DUAT.

Flight service station/FSS—Facilities which provide pilot briefing, enroute communications and VFR search and rescue services, assist lost aircraft and aircraft in emergency situations, relay ATC clearances, disseminate Notices to Airmen, broadcast aviation weather and NAS information, receive and process IFR flight plans, and monitor navaids. In addition, at selected locations, FSSes provide Enroute Flight Advisory Service (Flight Watch), take weather observations, issue airport advisories, and advise Customs and Immigration of transborder flights.

Flight standards district office/FSDO—An FAA field office serving an assigned geographical area and staffed with flight standards personnel who serve the aviation industry and the general public on matters relating to the certification and operation of air carrier and general aviation aircraft. Activities include general surveillance of operational safety, certification of airmen and aircraft, accident prevention, investigation, enforcement, etc.

Flight watch—A shortened term for use in air-ground contacts to identify the flight service station providing Enroute Flight Advisory Service; e.g., "Oakland Flight Watch."

Flow control—Measures designed to adjust the flow of traffic into a given airspace, along a given route, or bound for a given airport so as to ensure the most effective utilization of the airspace.

Gate hold procedures—Procedures at selected airports to hold aircraft at the gate or other ground location whenever departure delays exceed or are anticipated to exceed 15 minutes. The sequence for departure will be maintained in accordance with initial call-up unless modified by flow control restrictions. Pilots should monitor the ground control/clearance delivery frequency for engine startup advisories or new proposed start time if the delay changes.

Glideslope/glidepath intercept altitude—The minimum altitude to intercept the glideslope/path on a precision approach.

Go around—Instructions for a pilot to abandon his approach to landing. Additional instructions may follow. Unless otherwise advised by ATC, a VFR aircraft or an aircraft conducting a visual approach should overfly the runway while climbing to traffic pattern altitude and enter the traffic pattern via the crosswind leg. A pilot on an IFR flight plan making an instrument approach should execute the published missed approach procedure or proceed as instructed by ATC.

Ground clutter—A pattern produced on the radar scope by ground returns which may degrade other radar returns in the affected area. The effect of ground clutter is minimized by the use of moving target indicator (MTI) circuits in the radar equipment resulting in a radar presentation which displays only targets which are in motion.

Ground controlled approach/GCA—A radar approach system operated from the ground by air traffic control personnel transmitting instructions to the pilot by radio. The approach may be conducted with surveillance radar (ASR) only or with both surveillance and precision approach radar (PAR). Usage of the term "GCA" by pilots is discouraged except when referring to a GCA facility. Pilots should specifically request a "PAR" approach when a precision radar approach is desired or request as "ASR" or "surveillance" approach when a non-precision radar approach is desired.

Ground delay—The amount of delay attributed to ATC, encountered prior to departure, usually associated with a CDT program.

Handoff—An action taken to transfer the radar identification of an aircraft from one controller to another if the aircraft will enter the receiving controller's airspace and radio communications with the aircraft will be transferred.

High speed taxiway/exit/turnoff—A long radius taxiway designed and provided with lighting or marking to define the path of aircraft, traveling at high speed (up to 60 knots), from the runway center to a point on the center of the taxiway. Also referred to as a long radius exit or turn-off taxiway. The high speed taxiway is designed to expedite aircraft turning off the runway after landing, thus reducing runway occupancy time.

Hold/holding procedure—A predetermined maneuver which keeps aircraft within a specified airspace while awaiting further clearance from air traffic control. Also used during ground operations to keep aircraft within a specified area or at a specified point while awaiting further clearance from air traffic control.

Holding fix—A specified fix identifiable to a pilot by navaids or visual reference to the ground used as a reference point in establishing and maintaining the position of an aircraft while holding.

Hold for release—Used by ATC to delay an aircraft for traffic management reasons; e.g., weather, traffic volume, etc. Holds for release instructions (including departure delay information) are used to inform a pilot or a controller (either directly or though an authorized relay) that a departure clearance is not valid until a release time or additional instructions have been received.

Ident—A request for a pilot to activate the aircraft transponder identification feature. This will help the controller to identify the aircraft on the radar screen.

IFR military training route (IR)—Routes used by the Department of Defense and associated Reserve and Air Guard units for the purpose of conducting low-altitude navigation and tactical training in both IFR and VFR weather conditions below 10,000 feet MSL at airspeeds in excess of 250 knots IAS.

Instrument approach procedure/IAP/instrument approach—A series of predetermined maneuvers for the orderly transfer of an aircraft under instrument flight conditions from the beginning of the initial approach to a landing or to a point from which a landing may be made visually.

Instrument flight rules/IFR—Rules governing the procedures for conducting instrument flight. Also a term used by pilots and controllers to indicate type of flight plan.

Instrument landing system/ILS—A precision instrument approach system which normally consists of the following electronic components and visual aids:

- Localizer
- Glideslope
- Outer Marker
- Middle Marker
- Approach Lights

Instrument meteorological conditions/IMC—Meteorological conditions expressed in terms of visibility, distance from cloud, and ceiling less than the minimums specified for visual meteorological conditions.

Instrument runway—A runway equipped with electronic and visual navigation aids for which a precision or non-precision approach procedure having straight-in landing minimums has been approved

Interrogator—The ground-based surveillance radar beacon transmitter-receiver, which normally scans in synchronism with a primary radar, transmitting discrete radio signals which repeatedly request all transponders on the mode being used to reply. The replies received are mixed with the primary radar returns and displayed on the same plan position indicator (radar scope). Also, applied to the airborne element of the TACAN/DME system.

Intersection—A point defined by any combination of courses, radials, or bearings of two or more navigational aids. Also used to describe the point where two runways, a runway and taxiway, or two taxiways cross or meet.

Jet route—A route designed to serve aircraft operations from 18,000 feet MSL up to and including flight level 450. The routes are referred to as "J" routes with numbering to identify the designated route; e.g. J105.

Known traffic—With respect to ATC clearances, means aircraft whose altitude, position, and intentions are known to ATC.

Lateral separation—The lateral spacing of aircraft at the same altitude by requiring operation on different routes or in different geographical locations.

Localizer—The component of an ILS which provides course guidance to the runway.

Localizer type directional aid/LDA—A navaid used for nonprecision instrument approaches with utility and accuracy comparable to a localizer but which is not part of a complete ILS and is not aligned with the runway.

Local traffic—Aircraft operating in the traffic pattern or within sight of the tower, or aircraft known to be departing or arriving from flight in local practice areas, or aircraft executing practice instrument approaches at the airport.

Low altitude airway structure/federal airways—The network of airways serving aircraft operations up to but not including 18,000 feet MSL.

Low altitude alert system/LAAS—An automated function of ATC equipment that alerts the controller when a Mode C transponder equipped aircraft on an IFR flight plan is below a predetermined minimum safe altitude. If requested by the pilot, LAAS monitoring is also available to VFR Mode C transponder-equipped aircraft.

Low approach—An approach over an airport or runway following an instrument approach or a VFR approach including the go-around maneuver where the pilot intentionally does not make contact with the runway.

Make short approach—Used by ATC to inform a pilot to alter his traffic pattern so as to make a short final approach.

Mandatory altitude—An altitude depicted on an instrument Approach Procedure Chart requiring the aircraft to maintain altitude at the depicted value.

Marker beacon—A part of ILS approaches, it marks specific points on the glideslope. Typically the outer marker will mark the point at which the glideslope is intercepted, while the inner marker marks the missed approach point.

Mayday—The international radiotelephony distress signal. When repeated three times, it indicates imminent and grave danger and that immediate assistance is requested.

Metering—A method of time-regulating arrival traffic flow into a terminal area so as not to exceed a predetermined terminal acceptance rate.

Metering fix—A fix along an established route from over which aircraft will be metered prior to entering terminal airspace. Normally, this fix should be established at a distance from the airport which will facilitate a profile descent 10,000 feet above airport elevation (AAE) or above.

Military training routes/MTR—Airspace of defined vertical and lateral dimensions established for the conduct of military night training at airspeeds in excess of 250 knots IAS.

Minimum IFR altitudes/MIA—Minimum altitudes for IFR operations as prescribed in FAR part 91. These altitudes are published on aeronautical charts and prescribed in FAR part 95 for airways and routes, and in FAR part 97 for standard instrument approach procedures. If no applicable minimum altitude is prescribed in FAR parts 95 or 97, the following minimum IFR altitude applies:

> • In designated mountainous areas, 2,000 feet above the highest obstacle within a horizontal distance of 5 statute miles from the course to be flown; or
> • Other than mountainous areas, 1,000 feet above the highest obstacle within a horizontal distance of 5 statute miles from the course to be flown; or
> • As otherwise authorized by the Administrator or assigned by ATC.

Minimum obstruction clearance altitude/MOCA—The lowest published altitude in effect between radio fixes on VOR airways, off-airway routes, or route segments which meets obstacle clearance requirements for the entire route segment and which assures acceptable navigational signal coverage only within 25 statute (22 nautical) miles of a VOR.

Minimum vectoring altitude/MVA—The lowest MSL altitude at which an IFR aircraft will be vectored by a radar controller, except as otherwise authorized for radar approaches, departures, and missed approaches. The altitude meets IFR obstacle clearance criteria. It may be lower than the published MEA along an airway or J-route segment. It may be utilized for radar vectoring only upon the controller's determination that an adequate radar return is being received from the aircraft being controlled. Charts depicting minimum vectoring altitudes are normally available only to the controllers and not to pilots.

Missed approach—

- A maneuver conducted by a pilot when an instrument approach cannot be completed to a landing. The route of flight and altitude are shown on instrument approach procedure charts. A pilot executing a missed approach prior to the Missed Approach Point (MAP) must continue along the final approach course to the MAP. The pilot may climb immediately to the altitude specified in the missed approach procedure.
- A term used by the pilot to inform ATC that he is executing the missed approach.
- At locations where ATC radar service is provided, the pilot should conform to radar vectors when provided by ATC in lieu of the published missed approach procedure.

Mode—The letter or number assigned to a specific pulse spacing of radio signals transmitted or received by ground interrogator or airborne transponder components of the Air Traffic Control Radar Beacon System (ATCRBS). Mode A (military Mode 3) and Mode C (altitude reporting) are used in air traffic control.

Movement area—The runway, taxiways, and other areas of an airport which are utilized for taxiing, takeoff, and landing of aircraft, exclusive of loading ramps and parking areas. At those airports with a tower, specific approval for entry onto the movement area must be obtained from ATC.

NAS stage A—The enroute ATC system's radar, computers and computer programs, controller plan view displays (PVDs/radarscopes), input/output devices, and the related communications equipment which are integrated to form the heart of the automated IFR air traffic control system. This equipment performs Flight Data Processing (FDP) and Radar Data Processing (RDP).

It interfaces with automated terminal systems and is used in the control of en route IFR aircraft.

National Airspace System/NAS—The common network of U.S. airspace; air navigation facilities, equipment and services, airports or landing areas; aeronautical charts, information and services; rules, regulations and procedures, technical information, and manpower and material. Included are system components shared jointly with the military.

National beacon code allocation plan airspace/NBCAP airspace—Airspace over United States territory located within the North American continent between Canada and Mexico, including adjacent territorial waters outward to boundaries of oceanic control areas (CTA)/flight information regions (FIR).

National Flight Data Center/NFDC—A facility in Washington D.C., established by FAA to operate a central aeronautical information service for the collection, validation, and dissemination of aeronautical data in support of the activities of government, industry and the aviation community. The information is published in the National Flight Data Digest.

Navaid classes—VOR, VORTAC, and TACAN aids are classed according to their operational use. The three classes of navaids are:

    T-Terminal
    L-Low altitude
    H-High altitude

The normal service range for T,L, and H class aids is found in the AIM. Certain operational requirements make it necessary to use some of these aids at greater service ranges than specified. Extended range is made possible through flight inspection determinations. Some aids also have lesser service range due to location, terrain, frequency protection, etc. Restrictions to service range are listed in Airport/Facility Directory.

Negative contact—Used by pilots to inform ATC that:

• Previously issued traffic is not in sight. It may be followed by the pilot's request for the controller to provide assistance in avoiding the traffic.
• They were unable to contact ATC on a particular frequency.

Nonapproach control tower—Authorizes aircraft to land or take off at the airport controlled by the tower or to transit the airport traffic area. The primary function of a nonapproach control tower is the sequencing of aircraft in the traffic pattern and on the landing area. Nonapproach control towers also separate aircraft operating under IFR clearances. They provide ground control services to aircraft, vehicles, personnel, and equipment on the airport movement area.

Numerous targets vicinity (location)—A traffic advisory issued by ATC to advise pilots that targets on the radar scope are too numerous to issue individual advisories for.

Option approach—An approach requested and conducted by a pilot which will result in either a touch-and-go, missed approach, low approach, stop-and-go, or full stop landing.

Outer area (associated with ARSA)—Nonregulatory airspace surrounding designated ARSA airports wherein ATC provides radar vectoring and sequencing on a full-time basis for all IFR and participating VFR aircraft. The service provided in the outer area is called ARSA service which includes: IFR/IFR-standard IFR separation; IFR/VFR-traffic advisories and conflict resolution; and VFR/VFR-traffic advisories and, as appropriate, safety alerts. The normal radius is 20 nautical miles with some variations based on site-specific requirements. The outer area extends outward from the primary ARSA airport and extends from the lower limits of radar/radio coverage up to the ceiling of the approach control's delegated airspace excluding the ARSA and other airspace as appropriate.

Pan-pan—The international radio-telephony urgency signal. When repeated three times, indicates uncertainty or alert followed by the nature of the urgency.

Parallel ILS/MLS approaches—Approaches to parallel runways by IFR aircraft which, when established inbound toward the airport on the adjacent final approach courses, are radar-separated by at least 2 miles.

Pilot briefing/preflight pilot briefing—A service provided by the FSS to assist pilots in flight planning. Briefing items may include weather information, notams, military activities, flow control information, and other items as requested.

Pilot automatic telephone weather answering service/ PATWAS—A continuous telephone recording containing current and forecast weather information for pilots.

Pilot's discretion—When used in conjunction with altitude assignments, means that ATC has offered the pilot the option of starting climb or descent whenever he wishes and conducting the climb or descent at any rate he wishes. He may temporarily level off at any intermediate altitude. However, once he has vacated an altitude, he may not return to that altitude.

Positive control—The separation of all air traffic within designated airspace by air traffic control.

Preferential route—Preferential routes (PDRs, PARs, and PDARs) are programmed into ARTCC computers to accomplish inter/intrafacility controller coordination and to assure that flight data is posted at the proper control positions, Locations having a need for these specific inbound and outbound routes normally publish such routes in local facility bulletins, and their use by pilots minimizes flight plan route amendments. When the work load or traffic situation permits, controllers normally provide radar vectors or assign requested routes to minimize circuitous routing. Preferential routes are usually confined to one ARTCC's area and are referred to by the following names or acronyms:

- Preferential Departure Route/PDR—A specific departure route from an airport or terminal area to an en route point where there is no further need for flow control. It may be included in a Standard Instrument Departure (SID) or a Preferred IFR route.
- Preferential Arrival Route/PAR—A specific arrival route from an appropriate en route point to an airport or terminal area. It may be included in a Standard Terminal Arrival (STAR) or Preferred IFR Route. The abbreviation "PAR" is used primarily within the ARTCC and should not be confused with the abbreviation for Precision Approach Radar.
- Preferential Departure and Arrival Route/PDAR—A route between two terminals which are within or immediately adjacent to one ARTCC's area. PDARs are not synonymous with Preferred IFR Routes but may be listed as such as they do accomplish essentially the same purpose.

Preferred IFR routes—Routes established between busier airports to increase system efficiency and capacity. They normally extend through one or more ARTCC areas and are designed to achieve balanced traffic flows among high density terminals. IFR clearances are issued on the basis of these routes except when severe weather avoidance procedures or other factors dictate otherwise, Preferred IFR routes are listed in the Airport/Facility Directory. If a flight is planned to or from an area having such routes but the departure or arrival point is not listed in the Airport/Facility Directory, pilots may use that part of a preferred IFR route which is appropriate for the departure or arrival point that is listed. Preferred IFR routes are correlated with SIDs and STARs and may be defined by airways, Jet routes, direct routes between navaids, waypoints, navaid radials/DME, or any combinations thereof.

Profile descent—An uninterrupted descent (except where level flight is required for speed adjustment; e.g., 250 knots at 10,000 feet MSL) from cruising altitude/level to interception of a glide slope or to minimum altitude specified for the initial or intermediate approach segment of a non-precision instrument approach. The profile descent normally terminates at the approach gate or where the glide slope or other appropriate minimum altitude is intercepted.

Published route—A route for which an IFR altitude has been established and published; e.g., Federal Airways, Jet Routes, Area Navigation Routes, Specified Direct Routes.

Quick look—A feature of NAS Stage A and ARTS which provides the controller the capability to display full data blocks of tracked aircraft from other control positions.

Quota flow control/QFLOW—A flow control procedure by which the Central Flow Control Function (CFCF) restricts traffic to the ARTC Center area having an impacted airport, thereby avoiding sector/area saturation.

Radar (RAdio Detection And Ranging)—A device which, by measuring the time interval between transmission and reception of radio pulses and correlating the angular orientation of the radiated antenna beam or beams in azimuth and/or elevation, provides information on range, azimuth, and/or elevation of objects in the path of the transmitted pulses.

Radar approach control facility—A terminal ATC facility that uses radar and nonradar capabilities to provide approach control services to aircraft arriving, departing, or transiting airspace controlled by the facility. Provides radar ATC services to aircraft operating in the vicinity of one or more civil and/or military airports in a terminal area. The facility may provide services of a ground controlled approach (GCA); i.e., ASR and PAR approaches. A radar approach control facility may be operated by FAA, USAF, US Army, USN, USMC, or jointly by FAA and a military service. Specific facility nomenclatures are used for administrative purposes only and are related to the physical location of the facility and the operating service generally as follows:

- Army Radar Approach Control/ARAC (Army)
- Radar Air Traffic Control Facility/RATCF (Navy/FAA)
- Terminal Radar Approach Control/TRACON (FAA)
- Tower/Airport Traffic Control Tower/ATCT (FAA). (Only those towers delegated approach control authority.)

Radar contact—

- Used by ATC to infirm an aircraft that it is identified on the radar display and radar flight following will be provided until radar identification is terminated. Radar service may also be provided within the limits of necessity and capability. When a pilot is informed of "radar contact," he automatically discontinues reporting over compulsory reporting points.
- The term used by one controller to inform another that the aircraft is identified and approval is granted for the aircraft to enter the receiving controller's airspace.

Radar contact lost—Used by ATC to inform a pilot that radar identification of his aircraft has been lost. The loss may be attributed to several things including the aircraft's merging with weather or ground clutter, the aircraft's flying below radar line of sight, the aircraft's entering an area of poor radar return, or a failure of the aircraft transponder or the ground radar equipment.

Radar flight following—The observation of the progress of radar identified aircraft, whose primary navigation is being provided by the pilot, wherein the controller retains and correlates the aircraft identity with the appropriate target or target symbol displayed on the radar scope.

Radar service terminated—Used by ATC to inform a pilot that he will no longer be provided any of the services that could be received while in radar contact. Radar service is automatically terminated, and the pilot is not advised in the following cases:

• An aircraft cancels its IFR flight plan, except within a TCA, TRSA, ARSA, or where Stage II service is provided.
• An aircraft conducting an instrument, visual, or contact approach has landed or has been instructed to change to advisory frequency.
• An arriving VFR aircraft, receiving radar service to a tower-controlled airport within a TCA, TRSA, ARSA, or where Stage II service is provided, has landed; or to all other airports, when instructed to change to tower or advisory frequency.
• An aircraft completes a radar approach.

Radar traffic advisories—Advisories issued to alert pilots to known or observed radar traffic which may affect the intended route of flight of their aircraft.

Radial—A magnetic bearing extending from a VOR/VORTAC/TACAN navigation facility.

Receiving controller/facility—A controller/facility receiving control of an aircraft from another controller/facility.

Release time—A departure time restriction issued to a pilot by ATC (either directly or through an authorized relay) when necessary to separate a departing aircraft from other traffic.

Request full route clearance/FRC—Used by pilots to request that the entire route of flight be read verbatim in an ATC clearance. Such request should be made to preclude receiving an ATC clearance based on the original filed flight plan when a filed IFR flight plan has been revised by the pilot, company, or operations prior to departure.

Resume own navigation—Used by ATC to advise a pilot to resume his own navigational responsibility. It is issued after completion of a radar vector or when radar contact is lost while the aircraft is being radar vectored.

Roger—I have received all of your last transmission. It should not be used to answer a question requiring a yes or a no answer.

Runway heading—The magnetic direction indicated by the runway number. When cleared to "fly/maintain runway heading," pilots are expected to comply with the ATC clearance by flying the heading indicated by the runway number without applying any drift correction; e.g., Runway 4, 40 degree magnetic heading; Runway 20, 200 degree magnetic heading.

Runway profile descent—An instrument flight rules (IFR) air traffic control arrival procedure to a runway published for pilot use in graphic and/or textual form and may be associated with a STAR. Runway Profile Descents provide routing and may depict crossing altitudes, speed restrictions, and headings to be flown from the enroute structure to the point where the pilot will receive clearance to execute an instrument approach procedure.

Runway use program—A noise abatement runway selection plan designed to enhance noise abatement efforts with regard to airport communities for arriving and departing aircraft. These plans are developed into runway use programs and apply to all turbojet aircraft 12,500 pounds or heavier; turbojet aircraft less than 12,500 pounds are included only if the airport proprietor determines that the aircraft creates a noise problem. Runway use programs are coordinated with FAA offices, and safety criteria used in these programs are developed by the Office of Flight Operations. Runway use programs are administered by the Air Traffic Service as "Formal" or "Informal" programs.

> • Formal Runway Use Program—An approved noise abatement program which is defined and acknowledged in a Letter of Understanding between Flight Operations, Air Traffic Service, the airport proprietor, and the users. Once established, participation in the program is mandatory for aircraft operators and pilots.
> • Informal Runway Use Program—An approved noise abatement program which does not require a Letter of Understanding, and participation in the program is voluntary for aircraft operators/pilots.

Safety alert—A safety alert issued by ATC to aircraft under their control if ATC is aware the aircraft is at an altitude which, in the controller's judgment, places the aircraft in unsafe proximity to terrain, obstructions, or other aircraft. The controller may discontinue the issuance of further alerts if the pilot advises he is taking action to correct the situation or has the other aircraft in sight.

> • Terrain/Obstruction Alert—A safety alert issued by ATC to aircraft under their control if ATC is aware the aircraft is at an altitude which, in the controller's judgment, places the aircraft in unsafe proximity to terrain/obstructions; e.g., "Low Altitude Alert, check your altitude immediately."

• Aircraft Conflict Alert—A safety alert issued by ATC to aircraft under their control if ATC is aware of an aircraft that is not under their control at an altitude which, in the controller's judgment, places both aircraft in unsafe proximity to each other. With the alert, ATC will offer the pilot an alternate course of action when feasible; e.g., "Traffic Alert, advise you turn right heading zero niner zero or climb to eight thousand immediately."

The issuance of safety alert is contingent upon the capability of the controller to have an awareness of an unsafe condition. The course of action provided will be predicated on other traffic under ATC control. Once the alert is issued, it is solely the pilot's prerogative to determine what course of action, if any, he will take.

Say again—Used to request a repeat of the last transmission. Usually specifies a transmission or portion thereof not understood or received; e.g., "Say again all after...."

Say altitude—Used by ATC to ascertain an aircraft's specific altitude/flight level. When the aircraft is climbing or descending, the pilot should state the indicated altitude rounded to the nearest 100 feet.

Say heading—Used by ATC to request an aircraft heading. The pilot should state the actual heading of the aircraft.

See and avoid—A visual procedure wherein pilots of aircraft flying in visual meteorological conditions (VMC), regardless of type of flight plan, are charged with the responsibility to observe the presence of other aircraft and to maneuver their aircraft as required to avoid the other aircraft. Right-of-way rules are contained in FAR part 91.

Severe weather avoidance plan/SWAP—An approved plan to minimize the effect of severe weather on traffic flows in impacted terminal and/or ARTCC areas. SWAP is normally implemented to provide the least disruption to the ATC system when flight through portions of airspace is difficult or impossible due to severe weather.

Severe weather forecast alerts/AWW—Preliminary messages issued in order to alert users that a Severe Weather Watch Bulletin (WW) is being issued. These messages define areas of possible severe thunderstorms or tornado activity. The messages are unscheduled and issued as required by the National Severe Storm Forecast Center at Kansas City, Missouri.

Short range clearance—A clearance issued to a departing IFR flight which authorizes IFR flight to a specific fix short of the destination while air traffic control facilities are coordinating and obtaining the complete clearance.

Sidestep maneuver—A visual maneuver accomplished by a pilot at the completion of an instrument approach to permit a straight-in landing on a parallel runway not more than 1,200 feet to either side of the runway to which the approach was conducted.

Simultaneous ILS/MLS approaches—An approach system permitting simultaneous ILS/MLS approaches to airports having parallel runways separated by at least 4,300 feet between centerlines. Integral parts of a total system are ILS/MLS, radar, communications, ATC procedure, and appropriate airborne equipment.

Special emergency—A condition of air piracy or other hostile act by a person(s) aboard an aircraft which threatens the safety of the aircraft or its passengers.

Special use airspace—Airspace of defined dimensions identified by an area on the surface of the earth wherein activities must be confined because of their nature and/or wherein limitations may be imposed upon aircraft operations that are not a part of those activities. Types of special use airspace are:

> • Alert Area—Airspace which may contain a high volume of pilot training activities or an unusual type of aerial activity, neither of which is hazardous to aircraft. Alert Areas are depicted on aeronautical charts for the information of nonparticipating pilots. All activities within an Alert Area are conducted in accordance with Federal Aviation Regulations, and pilots of participating aircraft as well as pilots transiting the area are equally responsible for collision avoidance.
> • Controlled Firing Area—Airspace wherein activities are conducted under conditions so controlled as to eliminate hazards to nonparticipating aircraft and to ensure the safety of persons and property on the ground.
> • Military Operations Area (MOA)—An MOA is an airspace assignment of defined vertical and lateral dimensions established outside positive control areas to separate/segregate certain military activities from IFR traffic and to identify for VFR traffic where these activities are conducted.

- Prohibited Area—Airspace within which the flight of aircraft is prohibited.
- Restricted Area—Airspace within which the flight of aircraft, while not wholly prohibited, is subject to restriction. Most restricted areas are designated joint use and IFR/VFR operations in the area may be authorized by the controlling ATC facility when it is not being utilized by the using agency. Restricted areas are depicted on enroute charts. Where joint use is authorized, the name of the ATC controlling facility is also shown.
- Warning Area—Airspace which may contain hazards to nonparticipating aircraft in international airspace.

Special VFR conditions—Weather conditions in a control zone or Class D airspace which are less than basic VFR and in which some aircraft are permitted flight under Visual Flight Rules.

Special VFR operations—Aircraft operating in accordance with clearances within control zones in weather conditions less than the basic VFR weather minima. Such operations must be requested by the pilot and approved by ATC.

Speed adjustment—An ATC procedure used to request pilots to adjust aircraft speed to a specific value for the purpose of providing desired spacing. Pilots are expected to maintain a speed of plus or minus 10 knots or 0.02 mach number of the specified speed. Examples of speed adjustments are:

- "Increase/reduce speed to mach point (number)."
- "Increase/reduce speed to (speed in knots)" or "Increase/reduce speed (number of knots) knots."

Squawk—Activate specific modes/codes/functions on the aircraft transponder; e.g., "Squawk two one zero five."

Standard instrument departure/SID—A preplanned instrument flight rules (IFR) air traffic control departure procedure printed for pilot use in graphic and/or textual form, SIDs provide transition from the terminal to the appropriate enroute structure.

Standard terminal arrival/STAR—A preplanned instrument flight rules (IFR) air traffic control arrival procedure published for pilot use in graphic and/or textual form. STARs provide transition from the enroute structure to an outer fix or an instrument approach fix/arrival way point in the terminal area.

Stepdown fix—A fix permitting additional descent within a segment of an instrument approach procedure by identifying a point at which a controlling obstacle has been safely overflown.

Stop altitude squawk—Used by ATC to inform an aircraft to turn off the automatic altitude reporting feature of its transponder. It is issued when the verbally reported altitude varies 300 feet or more from the automatic altitude report.

Stop and go—A procedure wherein an aircraft will land, make a complete stop on the runway, and then commence a takeoff from that point.

Stopover flight plan—A flight plan format which permits in a single submission the filing of a sequence of flight plans through interim fill-stop destinations to a final destination.

Stop squawk—Used by ATC to tell the pilot to turn specified functions of the aircraft transponder off.

Straight-in approach-IFR—An instrument approach wherein final approach is begun without first having executed a procedure turn, not necessarily completed with a straight-in landing or made to straight-in landing minimums.

Straight-in approach-VFR—Entry into the traffic pattern by interception of the extended runway centerline (final approach course) without executing any other portion of the traffic pattern.

Straight-in landing—A landing made on a runway aligned within 30 degrees of the final approach course following completion of an instrument approach.

Surveillance approach—An instrument approach wherein the air traffic controller issues instructions, for pilot compliance, based on aircraft position in relation to the final approach course (azimuth), and the distance (range) from the end of the runway as displayed on the controller's radar scope. The controller will provide recommended altitudes on final approach if requested by the pilot.

Target symbol—A computer-generated indication shown on a radar display resulting from a primary radar return or a radar beacon reply.

Taxi—The movement of an airplane under its own power on the surface of an airport. Also, it describes the surface movement of helicopters equipped with wheels.

Taxi into position and hold—Used by ATC to inform a pilot to taxi onto the departure runway in takeoff position and hold. It is not authorization for takeoff. It is used when takeoff clearance cannot immediately be issued because of traffic or other reasons.

Terminal area facility—A facility providing air traffic control service for arriving and departing IFR, VFR, Special VFR, and on occasion enroute aircraft.

Terminal radar program—A national program instituted to extend the terminal radar services provided for IFR aircraft to VFR aircraft. Pilot participation in the program is urged but is not mandatory. The program is divided into two parts and referred to as Stage II and Stage III. The Stage service provided at a particular location is contained in the Airport/Facility Directory.

    • Stage I originally comprised two basic radar services (traffic advisories and limited vectoring to VFR aircraft). These services are provided by all commissioned terminal radar facilities, but the term "Stage I" has been deleted from use.

    • Stage II/Radar Advisory and Sequencing for VFR Aircraft. Provides, in addition to the basic radar services, vectoring and sequencing on a full-time basis to arriving VFR aircraft. The purpose is to adjust the flow of arriving IFR and VFR aircraft into the traffic pattern in a safe and orderly manner and to provide traffic advisories to departing VFR aircraft.

    • Stage III/Radar Sequencing and Separation Service for VFR Aircraft. Provides, in addition to the basic radar services and Stage II, separation between all participating VFR aircraft. The purpose is to provide separation between all participating VFR aircraft and all IFR aircraft operating within the airspace defined as a Terminal Radar Service Area (TRSA), Airport Radar Service Area (ARSA) or Terminal Control Area(TCA). (Class C and B airspace.)

Terminal radar service area/TRSA—Airspace surrounding designated airports wherein ATC provides radar vectoring, sequencing ,and separation on a full-time basis for all IFR and participating VFR aircraft. Service provided in a TRSA is called Stage III Service. **Note: This variety of airspace is virtually extinct, having been replaced in most instances by the ARSA. It will disappear entirely in 1993, being replaced by Class C airspace.**

Threshold—The beginning of the portion of the runway usable for landing.

Touchdown zone—The first 3,000 feet of the runway beginning at the threshold. The area is used for determination of Touchdown Zone Elevation in the development of straight-in landing minimums for instrument approaches.

Track—The actual flight path of an aircraft over the surface of the earth.

Traffic advisories—Advisories issued to alert pilots to other known or observed air traffic which may be in such proximity to the position or intended route of night of their aircraft to warrant their attention. Such advisories may be based on:

- Visual observation.
- Observation of radar identified and unidentified aircraft targets on an ATC radar display, or
- Verbal reports from pilots or other facilities.

The word "traffic" followed by additional information, if known, is used to provide such advisories; e.g., "Traffic, 2 o'clock, one zero miles, southbound, eight thousand." Traffic advisory service will be provided to the extent possible depending on higher priority duties of the controller or other limitations, e.g., radar limitations, volume of traffic, frequency congestion, or controller workload. Traffic advisories do not relieve the pilot of his responsibility to see and avoid other aircraft. Pilots are cautioned that there are many times when the controller is not able to give traffic advisories concerning all traffic in the aircraft's proximity; in other words, when a pilot requests or is receiving traffic advisories, he should not assume that all traffic will be called out.

Traffic alert and collision avoidance system/TCAS—An airborne collision avoidance system based on radar beacon signals which operates independent of ground-based equipment.

Traffic no longer a factor—Indicates that the traffic described in a previously issued traffic advisory is no longer of concern to the pilot.

Traffic pattern—The traffic flow that is prescribed for aircraft landing at, taxiing on, or taking off from an airport. The components of a typical traffic pattern are upwind leg, crosswind leg, downwind leg, base leg, and final approach.

- Upwind leg—A flight path parallel to the landing runway in the direction of landing.
- Crosswind leg—A flight path at right angles to the landing runway off its upwind end.

• Downwind Leg—A flight path parallel to the landing runway in the direction opposite to landing. The downwind leg normally extends between the crosswind leg and the base leg.

• Base Leg—A flight path at right angles to the landing runway off its approach end. The base leg normally extends from the downwind leg to the intersection of the extended runway centerline.

• Final Approach—A flight path in the direction of landing along the extended runway centerline. The final approach normally extends from the base leg to the runway. An aircraft making a straight-in approach VFR is also considered to be on final approach.

Transponder—The airborne radar beacon receiver/transmitter portion of the Air Traffic Control Radar Beacon System (ATCRBS) which automatically receives radio signals from interrogators on the ground, and selectively replies with a specific reply pulse or pulse group only to those interrogations being received on the mode to which it is set to respond.

Uncontrolled airspace—Uncontrolled airspace is that portion of the airspace that has not been designated as continental control area, control area, control zone, terminal control area, or transition area and within which ATC has neither the authority nor the responsibility for exercising control over air traffic. This will be designated as Class E airspace under the ICAO designation scheme.

Unicom—A nongovernment communication facility which may provide airport information at certain airports. Locations and frequencies of unicoms are shown on aeronautical charts and publications.

Unpublished route—A route for which no minimum altitude is published or charted for pilot use. It may include a direct route between navaids, a radial, a radar vector, or a final approach course beyond the segments of an instrument approach procedure.

Very high frequency/VHF—The frequency band between 30 and 300 mHz. Portions of this band, 108 to 118 mHz. are used for certain navaids; 118 to 136 mHz are used for civil air/ground voice communications. Other frequencies in this band are used for purposes not related to air traffic control.

VFR conditions—Weather conditions equal to or better than the minimum for flight under visual night rules. The term may be used as an ATC clearance/instruction only when:

- An IFR aircraft requests a climb/descent in VFR conditions.
- The clearance will result in noise abatement benefits where part of the IFR departure route does not conform to an FAA approved noise abatement route or altitude.
- A pilot has requested a practice instrument approach and is not on an IFR flight plan.

All pilots receiving this authorization must comply with the VFR visibility and distance from cloud criteria in FAR part 91. Use of the term does not relieve controllers of their responsibility to separate aircraft. When used as an ATC clearance/instruction, the term may be abbreviated.

VFR-on-top—ATC authorization for an IFR aircraft to operate in VFR conditions at any appropriate VFR altitude (as specified in the FARs and as restricted by ATC). A pilot receiving this authorization must comply with the VFR visibility distance from cloud criteria, and the minimum IFR altitudes specified in FAR part 91. The use of this term does not relieve controllers of their responsibility to separate aircraft.

VFR not recommended—An advisory provided by flight service stations to a pilot during a preflight or in-flight weather briefing that flight under visual flight rules is not recommended. To be given when the current and/or forecast weather conditions are at or below VFR minimums. It does not abrogate the pilot's authority to make his own decision.

Visual approach—An approach wherein an aircraft on an IFR flight plan, operating in VFR conditions under the control of an air traffic control facility and having an air traffic control authorization, may proceed to the airport of destination in VFR conditions.

Visual flight rules/VFR—Rules that govern the procedures for conducting flight under visual conditions. The term "VFR" is also used in the United States to indicate weather conditions that are equal to or greater than minimum VFR requirements. In addition, it is used by pilots and controllers to indicate type of flight plan.

Visual holding—The holding of aircraft at selected, prominent geographical fixes which can be easily recognized from the air.

Visual meteorological conditions/VMC—Meteorological conditions expressed in terms of visibility, distance from cloud, and ceiling equal to or better than specified minimums.

Visual separation—A means employed by ATC to separate aircraft in terminal areas. There are two ways to effect this separation:

- The tower controller sees the aircraft involved and issues instructions. as necessary, to ensure that the aircraft avoid each other.
- A pilot sees the other aircraft involved and upon instructions from the controller provides his own separation by maneuvering his aircraft as necessary to avoid it. This may involve following another aircraft or keeping it in sight until it is no longer a factor.

Vortices/wing tip vortices—Circular patterns of air created by the movement of an airfoil through the air when generating lift. As an airfoil moves through the atmosphere in sustained flight, an area of low pressure is created above it. The air flowing from the high pressure area to the low pressure area around and about the tips of the airfoil tends to roll up into two rapidly rotating vortices, cylindrical in shape. These vortices are the most predominant parts of aircraft wake turbulence and their rotational force is dependent upon the wing loading, gross weight, and speed of the generating aircraft. The vortices from medium to heavy aircraft can be of extremely high velocity and hazardous to smaller aircraft.

VOR/very high frequency omnidirectional range station—A ground-based electronic navigation aid transmitting very high frequency navigation signals, 360 degrees in azimuth, oriented from magnetic north. Used as the basis for navigation in the National Airspace System. The VOR periodically identifies itself by Morse Code and may have an additional voice identification feature, Voice features may be used by ATC or FSS for transmitting instructions/information to pilots.

Wake turbulence—Phenomena resulting from the passage of an aircraft through the atmosphere. The term includes vortices, thrust stream turbulence, jet blast, jet wash, propeller wash, and rotor wash both on the ground and in the air.

Wilco—I have received your message, understand it, and will comply with it.

Index

W